Principles of Rock Deformation

Petrology and Structural Geology

A Series of Books

Principles of
Rock Deformation

Adolphe Nicolas

Université des Sciences et Techniques du Languedoc, Montpellier II

D. Reidel Publishing Company

A MEMBER OF THE KLUWER ACADEMIC PUBLISHERS GROUP

Dordrecht / Boston / Lancaster / Tokyo

Library of Congress Cataloging-in-Publication Data

Nicolas, A. (Adolphe), 1936–
    Principles of rock deformation.

    (Petrology and structural geology)
    Translation of: Principes de tectonique.
    Bibliography: p.
    Includes index.
    1. Rock deformation.   I. Mainprice, D. H.   II.   Title.   III.   Series.
QE604.N5413   1986   551.8   86–22059
ISBN 90–277–2368–0
ISBN 90–277–2369–9 (pbk)

Published by D. Reidel Publishing Company,
P.O. Box 17, 3300 AA Dordrecht, Holland.

Sold and distributed in the U.S.A. and Canada
by Kluwer Academic Publishers,
101 Philip Drive, Assinippi Park, Norwell, MA 02061, U.S.A.

In all other countries, sold and distributed
by Kluwer Academic Publishers Group,
P.O. Box 322, 3300 AH Dordrecht, Holland.

Originally published in 1984 by Masson under the title
Principes de Tectonique
Translated from the French by S. W. Morel
English text edited by D. H. Mainprice

Printed in the Netherlands.

Table of Contents

Foreword

Physicists attempt to reduce natural phenomena to their essential dimensions by means of simplification and approximation and to account for them by defining natural laws. Paradoxically, whilst there is a critical need in geology to reduce the overwhelming field information to its essentials, it often remains in an over-descriptive state. This prudent attitude of geologists is dictated by the nature of the subjects being considered, as it is often difficult to derive the significant parameters from the raw data. It also follows from the way that geological work is carried out. Geologists proceed, as in a police investigation, by trying to reconstruct past conditions and events from an analysis of the features preserved in rocks. In physics all knowledge is based on experiment but in the Earth Sciences experimental evidence is of very limited scope and is difficult to interpret. The geologist's cautious approach in accepting evidence gained by modelling and quantification is sometimes questionable when it is taken too far. It shuts out potentially fruitful lines of advance ; for instance when refusing order of magnitude calculations, it risks being drowned in anthropomorphic speculation. Happily nowadays, many more studies tend to separate and order the significant facts and are carried out with numerical constraints, which although they are approximate in nature, limit the range of hypotheses and thus give rise to new models.

Structural geology has not escaped entirely from this tendency towards an over-profusion of detailed descriptions of geological structures of only anecdotal interest. It is true that we are now witnessing a renewal largely founded upon experimentation.

I have tried to present this recent progress, at times taking the opposite course of giving a traditional presentation and limiting my arguments to ones that appear to me to correspond to the essential structures and mechanisms of deformation.

As a consequence, the reader will only find here an **analytical** presentation of the essential structures. There is only a limited number of tectonic situations in the nature, such as stretching or shortening, folding, diapirism and gravitational tectonics which following their own laws combine essential structural elements. Study of these situations depends on structural **synthesis** and should be the object of another book.

Kinematic and dynamic study of natural structures sometimes requires one to simplify and approximate because the actual stage of development of these studies are that of first-order phenomena. Such fundamental issues like the origin of slaty cleavage in shear regime are still poorly understood. I have been rash, perhaps, in attempting to present them here.

Thanks to a more physical approach, one also hopes that the geological objects and processes will be better separated from their matrix thus achieving a simpler and more coherent layout. This approach does not require mathematical treatments and the work should be readily understood by everybody. Such treatments are presented in the Appendices.

I have consulted J.L. BOUCHEZ frequently during the writing of the book and it has also been reviewed critically by several colleagues, F. BOUDIER, J.P. BRUN and P. CHOUKROUNE who have also collaborated in the writing of Appendix II, P. COBBOLD, J.P. GRATIER, M. MATTAUER, D. MAINPRICE who also gave assistance in the english editing, J. MERCIER, A. PECHER, J.P. POIRIER, P.VIALON, J.L. VIGNERESSE and C. WILLAIME without forgetting our post-graduate students ; M. CANNAT, G. CEULENEER et S. SEROT-CRAMBERT. M.C. BREHIER and A. COSSARD were responsible for the printing and illustration of the work, and S.W. MOREL for the english translation. Thanks to them all.

Nantes, December 1983 and June 1986

A. NICOLAS

Chapter 1

Introduction

In the disciplines of earth sciences where the objective is not essentially an historical or applied one, it is possible to make two main divisions, with disciplines describing geological **structure** and others that are concerned with the **composition** of natural substances. Thus, crystallography, geophysics, tectonophysics, structural geology, tectonics, geomorphology, and in large measure, sedimentology, are concerned with natural structures and their evolution, whereas mineralogy, geochemistry and petrology are concerned with the composition and chemical evolution of natural substances. In geological studies it is necessary to combine these two aspects.

Within the web of the different earth sciences, structural geology can be approached from two directions. One is concerned with the origin and **history** of structures. In a deformed region such as in a mountain belt, the geometry of complex deformations, principally superimposed foldings, are investigated. One has to unravel the threads then, with the aid of stratigraphical, geochronological, petrological and geochemical methods, to elucidate the geological evolution of the region.

The **physical** approach (or **tectonophysical** which associates physical and geological considerations) deals with the detailed study of deformational mechanisms. Although the first theoretical and experimental work dates from the nineteenth century with the work of Sorby and Daubrée, the development of this approach is very recent. It is founded upon the considerable progress that has been made in the experimental and theoretical study of materials since the second world war. Similar studies on rock-forming minerals did not really begin until the sixties.

Beyond a **geometrical** description which permits deformation to be quantified, such studies can achieve results concerning the movements of rocks masses (**kinematics**) and the forces which produced them (**dynamics**). Integrated into a more comprehensive geophysical framework, these results lead to a better understanding of the geodynamics of the system or domain being studied. Thus in a seismic area, a better understanding of the local stress conditions and of the tectonic evolution may be gained by kinematic and dynamic analysis of faults (tectonophysics), combined with a study of processes at the seismic foci (seismology).

These two approaches, historical and physical, are not mutually exclusive. It is necessary to regard them as being complementary. Analysing the history of a mountain chain requires a description of historical as well as geodynamic conditions.

Geometrical, kinematic and dynamic aspects

There are three approaches to structural analysis, geometrical or structural (sensu stricto), kinematic and dynamic. When considering an object that has been deformed naturally, we are first of all obliged to make a geometrical description of it. If the object can be restored to its original shape, i.e. the shape it had before the deformation, it is possible to describe and quantify the amount of the strain that has taken place. This is **finite strain** analysis, that is to say, the total amount of strain that the object has undergone. This relies on structural geometrical analysis. We can also try to understand the various ways by which the object under consideration passed from its initial to its final state ; this is **kinematic analysis**. The final objective, **dynamic analysis**, aims to define the forces that are responsible for deformational processes. One imagines that, in the study of natural deformation, difficulties increase on passing successively from structural analysis to kinematic analysis and then to dynamic analysis. This is where deformation experiments are very helpful as the system of applied forces is known and it is easy to compare the initial and final states of the object being deformed. Deformation experiments also aim at understanding the **rheology** of the material concerned, that is how rapidly it deforms under varying applied conditions (stress, temperature, etc...).

Finally, we have favoured the **study of large homogeneous deformation** rather than heterogeneous deformation, mainly when the latter is modest as in the case of open folds. Two arguments justify this choice : the large displacements, overthrusts, stretching and shearing on the boundaries of plates or within their interiors are often localised into quite narrow zones where they are expressed by relatively homogeneous deformations of large amplitude. On the other hand it is precisely in the case of such deformations that one is now able to draw a few conclusions about the kinematics, dynamics and rheology thanks to the approach developed here.

Chapter 2

Strain and Stress

2.1. STRAIN

2.1.1. Definitions

When rocks are subjected to exterior forces they are deformed and suffer displacements, **translations and rotations**. Deformation, in a narrow sense, is the **change in form** undergone by the
rock body.

In translation the vectoral movements on different parts of
the rock are the same, whereas in deformation they vary from one
point to another, that is to say, there are gradients of displacement in the body of the rock (fig.2.1). Mathematically, the
strain is expressed with the help of tensors (see appendix I).

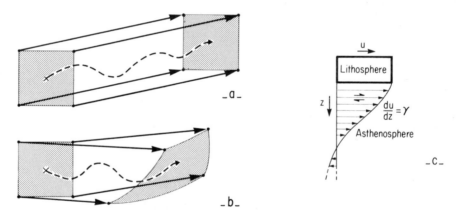

Fig.2.1. Displacement vectors in translation (a) and deformation
(b). In translation the displacement vectors are equal but in
deformation they are unequal. The actual path followed by a
particle during deformation (dashed line) does not necessarily
coincide with its displacement vector which relates only its
original and final positions. c) Application to upper mantle
motion close to an oceanic ridge. The undeforming lithosphere is
translated by equal vectors u . The deforming asthenosphere
undergoes a displacement gradient du/dz = γ , which is the shear
strain.

The determination of translations and rotations undergone in
relation to the geographic reference frame remains an important
geological objective. Thus one may want to measure the amount of

3

displacement of a nappe in a mountain belt or the amount of rotation about a vertical axis between two plates relative to each other (e.g. the rotation of the Corsica-Sardinian block in relation to Europe). Two methods allow us to arrive at a solution : the first employs a comparison between the initial and final states when the former is known (the source of nappes in the first example or the early paleomagnetic location in the second) ; the other method relies on integrating the deformations undergone by the formations being considered such as folds, shear zones etc. In chapter 7 we shall see how to use these ideas.

In the case of figure 2.6, it can be said that deformation is **heterogeneous** whereas in that of figure 2.7 it is **homogeneous**. More precisely, homogeneous deformation changes all straight lines in the solid into new straight lines (fig.2.2a) whereas in **heterogeneous** deformation at least some of these straight lines are changed into curves (fig.2.2b). In the case of heterogeneous deformation, the relative displacement of points within the material may be **continuous** or **discontinuous**. The discontinuity is expressed by a rupture, for example a fault (fig.2.2.c). A deformation that appears to be continuous and homogeneous when viewed on the scale of a massif or outcrop may be reduced at hand specimen scale or at that of the thin section to an accumulation of deformations on narrowly spaced and regular discontinuities. The deformation is called **penetrative** at the scale of an outcrop or massif(fig.2.3a) and **non-penetrative** at that of a hand specimen or a thin section (fig.2.3b).

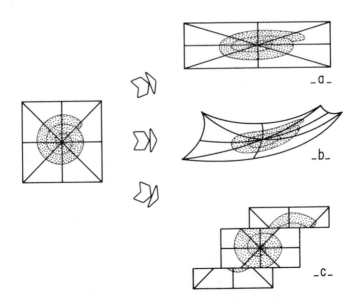

Fig.2.2. a) Continuous homogeneous deformation : straight lines deform into straight lines. b) Continuous heterogeneous deformation : a straight line is generally deformed into a curve. c) Discontinuous deformation.

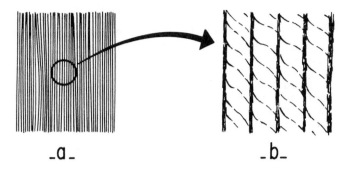

Fig.2.3. Deformation which is penetrative at outcrop scale (a)
can be non-penetrative at the scale of a hand specimen (b).

2.1.2. Finite strain ellipsoid

Let us imagine that the object is originally made up of
small spheres such as ooliths. After homogeneous deformation,
these spheres are deformed into ellipsoids and one can measure
their strain by comparing their shape and size with those of the
original spheres. This ellipsoid is called the **finite strain
ellipsoid**, the principal axes of which are X,Y,Z (see also
§2.2.2) (fig.2.4). In comparison between the ellipsoid and the
sphere that presumably preceded it, one must remember that
constant volume deformation can only be explicity assumed in the
case of plastic deformation sensu-stricto. As we shall see,
deformation in the presence of fluids is often accompanied by
loss of volume due to the expulsion of fluid.

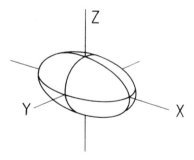

Fig.2.4. Strain ellipsoid with principal axes $X \geqslant Y \geqslant Z$.

In the case of homogeneous deformation, one measures the
strain of various markers (veins, pebbles, oxidation or reduction
spots, etc.) by a statistical estimate of the lengths of the
principal axes of the strain ellipsoid in the planes XY, XZ and
YZ. In the case of heterogeneous deformation the domain under
study is sub-divided into sub-domains in which deformation can be
regarded as homogeneous (§8.1). Measurement of strain has become
sophisticated thanks to the judicious use of many different
strain markers. A more complete analysis of this problem is
proposed in Appendix II.

2.1.3. States of strain

The state of strain is depicted by the shape of the strain ellipsoid. For example flattened or elongated strain ellipsoids represent different states. When the Y axis of the ellipsoid remains unchanged at constant volume, the strain is **plane**. The different states are represented very well on Flinn's diagram (fig.2.5). K represents the slope of a line from the data point to the origin at (1,1), so that :

K = a-1/b-1 with a = X/Y and b =Y/Z

K on the diagram defines several domains. Thus when K = 0, the finite strain ellipsoid is uniaxial oblate and flattened perpendicular to Z (such as in a pancake). Where 0<K<1, the strain ellipsoid is no longer uniaxial but remains oblate and of a flattening type. When 1<K<∞ , the strain ellipsoid is prolate and the strain is constrictive in character. When K = ∞ , the strain ellipsoid is uniaxial prolate or "cigar shaped", now stretched along X.

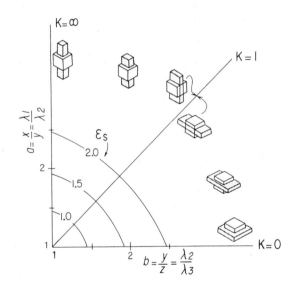

Fig.2.5. Diagram illustrating the differing states of finite strain and their relationships to the principal axes of the strain ellipsoid (Flinn's diagram). The little drawings illustrate the type of deformation in each domain. For the meaning of λ and ε s, see appendix II (Hobbs, Means and Williams, 1976. Wiley, New York).

2.2. ELEMENTS OF RHEOLOGY

2.2.1. Experimental deformation

There are two ways to undertake experiments on deformation.

The first way is by attempting to reproduce in the laboratory situations analogous to complex natural ones such as those of diapirism, convection and folding (fig.2.6). One is then forced to reduce the effects of space and time by a careful choice of weakly viscous fluids. The other way, with which we are concerned here, is to carry out mechanical tests on rocks under conditions that are as uniform as is possible. For example a cube is changed into an oblique prism (fig.2.7) or a cylinder is transformed into another cylinder of different diameter and length. These tests are devised to discover, by applying simple forces, how rapidly a crystalline sample deforms under different conditions of confining pressure, partial pressure, temperature etc. which are controlled in as precise a manner as is possible. As well as the response of the material which is called its **rheology**, one can examine the modifications to the form and internal structure of the sample during deformation and thus determine the structural signature of a known deformation.

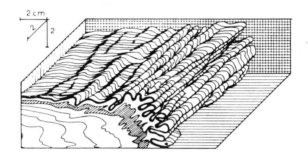

Fig.2.6. Experimental deformation of layers of plasticine (modelling clay) of varying viscosity by simple shear (Hugon and Cobbold, 1980. RAST, Soc. Geol. France, 186).

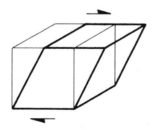

Fig.2.7. Experimental deformation of a right-angled crystalline prism.

2.2.2. Elastic, plastic and viscous deformation

A test sample of a single crystal or rock has a characteristic rheology when deformed under compression, as illustrated in figure 2.8a. In this figure, for each increment of pressure applied, also called the normal stress, (σ) (for a more complete

definition of stress, see Appendix I), the total amount of strain
is expressed by the corresponding amount of shortening :

e = (L1-Lo)/Lo where Lo = initial length, L1 = final length.

    This quantity is called **uniform finite strain**, as opposed to
the infinitesimal strain dL. It is not a geologically useful
measure of the strain. The **natural finite strain** εL is expressed
by :

$$\varepsilon_L = \int_{L_0}^{L_1} dL/L = \ln L_1/L_0$$

    For the moderate deformations that are generally achieved in
this type of test, equating the two quantities does not introduce
a significant error.

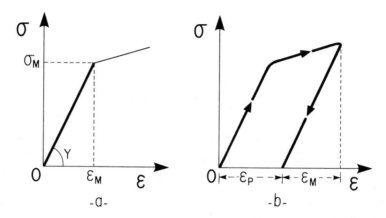

Fig.2.8. Deformation   of a rock under stress . In figure a the
thick line corresponds to the elastic domain. Figure b shows a
cycle with a plastic deformation ε p, beyond the elastic limit
(σ M, ε M).

    Figure 2.8 shows the response of a material by a line having
a steep slope and then a shallower slope. The first part corresponds generally with strains smaller than 1% and relationship
between applied stress and the resulting strain is expressed by :
σ =Y ε where Y = **Young's Modulus**. This linear behaviour corresponds to the domain of **elastic deformation**. Elastic deformation
is reversible, that is the sample reverts to its initial form if
the stress is removed. The transition point between this steep
line and the shallower line is called the **elastic limit** with
coordinates σ M and ε M on figure 2.8a. This corresponds to the
maximum stress that the sample can undergo without suffering
permanent strain.
    Above the elastic limit, one enters the domain of **plastic
deformation**, where if the applied stress is relaxed, only the
elastic part of the deformation is reversed, ε M = σM/Y
(fig.2.8b). The plastic deformation, shown by ε p on figure 2.8
is thus an irreversible change. The corresponding line on the
diagram is not necessarily a straight one. On figure 2.8 it has a

positive slope indicating that it is necessary to constantly increase the applied stress, showing that the material is **hardening**.

The slope of the line can be zero (σ p = const.). The flow under conditions of constant stress is called **creep** (fig.2.9). If the rate of deformation $\dot{\varepsilon}$ =d ε /dt is also constant, one speaks of **steady state flow**. It is under these conditions that the **flow of the material** is measured and expressed by the equation $\dot{\varepsilon}= A\,\sigma^{n}$; (T,P = const.). A is a constant and n,the exponent of the stress which depends on the flow mechanism.

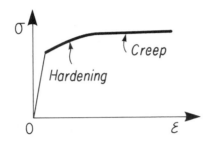

Fig.2.9. Plastic deformation (thick line) as a function of applied stress, beginning with hardening and followed by creep.

Liquids and certain solids can deform in a **viscous** manner. This mode of deformation is characterized by the presence of a linear relationship between the rate of deformation and the applied stress (for Newtonian fluids) : σ = $\eta\dot{\varepsilon}$. η is the **viscosity coefficient**. The rheology of viscous bodies is frequently marked by the absence of a threshold, like the elastic limit in the case of plastic deformation (fig.2.10). Natural fluids such as wet sediments saturated with water or magmas, may have a more complex behaviour with a viscosity that is no longer independent of stress.

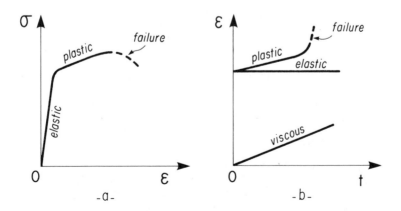

Fig.2.10. Elastic, plastic, viscous and brittle (ruptural) behaviour in diagrams (σ,ε) and (ε, t) ; σ = const. in diagram b.

Frequently, hardening accompanying plastic deformation (fig.2.9) is a precursor to the following stage of **rupture** or **fracture** (fig.2.10). This marks the transition from a **ductile** behaviour during the plastic stage to a **brittle** one at the onset of rupture. In tectonics where one is interested in permanent deformations one contrasts **continuous deformation**, (plastic or viscous ; ductile behaviour), with **discontinuous deformation** or **fracturing** (brittle behaviour). These definitions are phenomenological and may depend on the scale of observation employed (see §3.2.1).

2.2.3. Percolation thresholds : the viscous-plastic transition.

Certain processes lead to an interest in solid-liquid relationships and the transition between viscous and plastic behaviour, for example in the dehydration reactions in sedimentary or weakly metamorphosed rocks, the progressive melting of strongly metamorphosed rocks or the crystallization of magma. We shall now look at the case of partial melting.

At first, the pockets of liquid formed in a solid in the course of melting are isolated. For an ill-defined and probably variable proportion of liquid to crystals, the pockets of liquid tend to join up to form a continuous fluid film. This transition from a porous impermeable to a porous permeable state is known as the **first percolation threshold.** If melting continues, the remaining crystalline parts become increasingly isolated in a liquid matrix. At the point where the crystalline framework is no longer continuous, the **second percolation threshold** is reached. It corresponds to a proportion of liquid to crystals, probably variable and around 35 %. Beyond this new threshold the medium becomes a magma, that is to say, a suspension of crystals or clots of crystals in a liquid.

From the rheological point of view, the presence of a limited amount of liquid in the solid tends to weaken it, but the more spectacular change takes place when the second threshold is

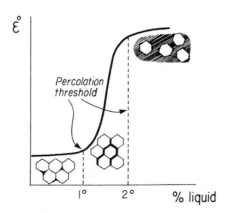

Fig.2.11. Increase in deformation rate $\overset{\circ}{\varepsilon}$ as a function of the liquid fraction in a solid-liquid mixture.

reached. This marks the transition from a generally plastic
regime in a solid framework, to a viscous regime of a suspension
of crystals in a liquid (fig.2.11). This transition certainly
plays a considerable role in the dynamics of magmas in the course
of crystallization. So, the motion of a granitic magma may stop
when its crystallization reaches this threshold, passing from
right to left on figure 2.11.

2.3. STRESS

2.3.1. Stress ellipsoid

      In the experiment shown in fig.2.12a, a test cylinder is
subjected to two opposed forces (F) applied perpendicularly to
opposite faces (S). Each force gives a pressure P = F/S, called
normal stress $\sigma 1$. In this case the applied stress is uniaxial. In
a more complex experiment the cylinder could be subjected to
distinct stresses, with $\sigma 1$ being perpendicular to the circular
sections and n at right angles to the sides of the cylinder
(fig.2.12b). In mechanics this is called a "triaxial" test. In
this experiment one can imagine that the sides of the cylinder
are put under a σ n uniform stress normal to the surface, whilst
an excess stress $\sigma = \sigma 1-\sigma n$ is exerted upon the circular cross-
section. This excess stress can induce deformation of the sample
and is called **differential stress** or **deviator**.

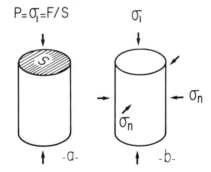

Fig.2.12. Deformation of a test cylinder under a) normal uniaxial
stress and b) triaxial stress.

      A more general situation can be imagined in which three
different stresses $\sigma 1 \geqslant \sigma 2 \geqslant \sigma 3$ are applied to the three opposed
faces of a rectangular brick or box (fig.2.13a). An **isotropic
pressure or mean stress** can then be defined as : Pi = 1/3
($\sigma 1+ \sigma 2+ \sigma 3$). This can be thought of as the hydrostatic component of the principal stresses and hence we can define three
principal differential stresses as, respectively $\sigma 1-$ Pi, $\sigma 2-$Pi,
$\sigma 3-$Pi. Instead of hydrostatic pressure, the terms **lithostatic or
geostatic pressure** are sometimes used to define the pressure
caused by the weight of superincumbent rocks and under experimental conditions, that of **confining pressure**.

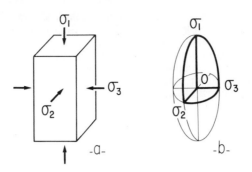

Fig.2.13.a) Test specimen with three stresses $\sigma 1 \geqslant \sigma 2 \geqslant \sigma 3$ applied perpendicularly to the faces. b) Stress ellipsoid corresponding to a).

The state of stress at point O is expressed mathematically by a tensor of six independent components (see Appendix I) and geometrically by an ellipsoid in which the three axes represent the three **normal principal stresses**, $\sigma 1$, $\sigma 2$ and $\sigma 3$ (fig.2.13b).

At a depth z near to the surface, the lithostatic pressure PL is orientated vertically and produced by the overlying rocks whose density ρ is written : $PL = \rho gz$, where g is the acceleration due to gravity.

The corresponding horizontal stresses in the absence of tectonic forces are : $\sigma 2 = \sigma 3 < PL$.

The lithostatic pressure PL becomes isotropic ($\sigma 1 = \sigma 2 = \sigma 3 = PL$) only at a depth of the order of 3 kilometers where all the shear stresses are relieved by deformation. This depth is where rocks begin to become ductile (fig.2.14).

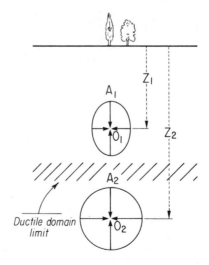

Fig.2.14. Change from lithostatic pressure at A1($\rho gz1$) to isostatic pressure at A2($\rho'gz2$) when the medium becomes ductile.

So far stresses that have been considered correspond to pressures, as they are perpendicular to the surface of the bodies. There are also stresses that are parallel to the surfaces, the **tangential or shearing stresses**. These are not pressures, although they are still defined by a force acting upon a surface. Let us consider the point X as an element of the surface S S' in any orientation with respect to a principal stress $\sigma 1$. This stress can be resolved into two components, one normal σ n1 and one tangential $\tau 1$ to this surface (fig.2.15). If one applies several principal stresses (for example $\sigma 1$ and $\sigma 3$ on figure 2.15), the sums of σ n1+σn3 = σ n and τ 1+τ3 = τ provide respectively the normal component σ n and tangential component τ at the point being considered.

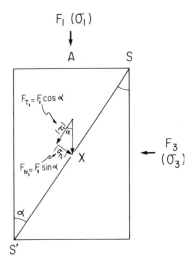

Fig.2.15. Breakdown of the force (F1) and the corresponding stresses operating at a point X of any plane SS', in a section perpendicular to σ 2.

2.3.2. The Mohr diagram

The shear stress and normal stress σ n components on a plane of known orientation with respect to the principal stresses can be quickly determined with the aid of a geometrical construction called **the Mohr diagram**.

Under the conditions illustrated in figure 2.13a, let us suppose for the sake of simplicity that $\sigma 2$ = ($\sigma 1$+ $\sigma 3$)/2, that is to say that $\sigma 2 = \sigma i$, the isotropic pressure. As a result the whole problem can be dealt with in the plane ($\sigma 1, \sigma 3$). Taking F1 as the force which acts perpendicularly on the surface A (fig.2.15), we look for the components of the forces at a point X on a surface SS' inclined at an angle α to the direction of F1. At the point X, the components of the force F1 are a normal force FN1 and a tangential force FT1 which, acting on the unit surface, give the following values to the normal stress σ n1 and the shear stress τ 1 :

$\sigma n1 = FN1/SS' = F1/A \sin^2\alpha = \sigma1 \dfrac{(1-\cos2\alpha)}{2}$

Likewise :

$\tau 1 = FT1/SS' = F1/A \cos\alpha \sin\alpha = \dfrac{\sigma1 \sin 2\alpha}{2}$

A similar resolution of forces enables one to know the contributions of the force F3 to σ n and τ for the portion of surface considered at point X. This leads to the following expressions :

$\sigma n = \sigma1\dfrac{(1-\cos2\alpha)}{2} + \sigma3 \dfrac{(1+\cos 2\alpha)}{2}$

or again :

$\sigma n = \dfrac{(\sigma1 + \sigma3)}{2} - \dfrac{(\sigma1 - \sigma3)}{2} \cos 2\alpha$

which, taking into account that $1/2(\sigma1+\sigma3)$ is equal to the isotropic pressure, the normal stress can be written :

$\sigma n = \sigma i - \dfrac{(\sigma1-\sigma3)}{2} \cos2\alpha$ \qquad (1)

Likewise :

$\tau = \dfrac{(\sigma1-\sigma3)}{2} \sin2\alpha$ \qquad (2)

The state of stress at point X on the surface SS' can now be represented on a diagram with the shear stress τ as the ordinate, and the normal stresses as the abscissa (fig.2.16). Equations (1) and (2) define a circle whose centre is at $\sigma i = 1/2(\sigma1+\sigma3)$ and

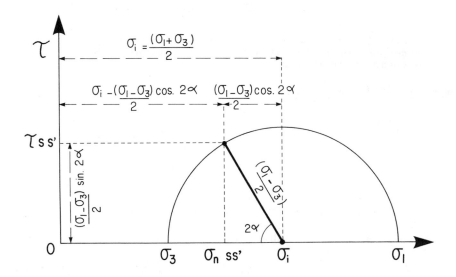

Fig.2.16. Mohr diagram reflecting the conditions of figure 2 15. Only the most significant (σ 1, σ 3) circle has been represented.

of radius = $1/2(\sigma 1 - \sigma 3)$. This is the **Mohr circle** and it is only
necessary to draw half of it. The principal stresses $\sigma 1$ and $\sigma 3$
lie on the abscissa at its point of intersection with the circle
and $\sigma 2 = 1/2 \ (\sigma 1 + \sigma 3)$ corresponds to its centre in accordance with
our initial choice ($\sigma 2 = \sigma i$). The values of the normal stress σ n
and shear stress τ acting upon a plane inclined at an angle α to
the principal stress direction are respectively the abscissa and
ordinate of the intersection of the circle with a line drawn from
the centre of the circle and making an angle of 2α with the axis
of the abcissae. Thus, τ is zero for $\alpha = 0°$ or $90°$, that is to
say along the faces parallel or perpendicular to $\sigma 1$ and reaches a
maximum value : $\tau M = 1/2 \ (\sigma 1 - \sigma 3)$ when $\alpha = 45°$. This analysis
demonstrates a remarkable character of the faces perpendicular to
the three principal stress directions, $\sigma 1$, $\sigma 2$, and $\sigma 3$: these are
submitted only to normal stresses.

2.4. PROGRESSIVE STRAIN

Finite strain transforming an initial sphere into an ellip-
soidal shape (the ellipsoid of finite strain or strain ellipsoid)
takes place by the addition of successive increments (see § 8.2) :
the strain is said to be **progressive**. For a very small increment

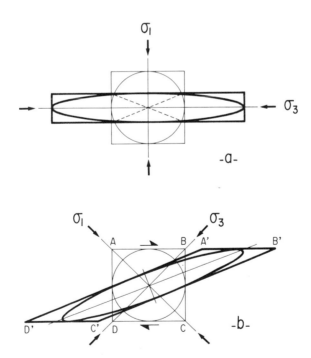

Fig.2.17. Regimes of deformation and stress-strain relations in
plane deformation, projected on the plane perpendicular to $\sigma 2$ and
the Y axis. a) Coaxial regime (pure shear) : dashed lines show
the invariant planes. b) Non-coaxial regime.

Chapter 3

Theory of Discontinuous Deformation

3.1. INTRODUCTION

Discontinuous deformation or rock fracturing has been well studied by experiments in the laboratory (in the field of Rock Mechanics) because of its numerous practical applications ; in civil engineering, rock bursts in quarrying and in mining, etc... New approaches have come from the study of hydraulic fracturing related to geothermal energy, to the in situ exploitation of combustible matter and the recovery of petroleum. Finally, a better understanding of the fundamental mechanism of rupture is needed in order to understand the processes of earthquakes, volcanic eruptions as well as the genesis of metallic mineral vein deposits.

In chapter 2 we have seen that rupture can occur in mechanical experiments at stresses above the elastic limit, after passing through a ductile state of variable extent (§2.2.2 and fig.2.10). The existence of a transition between ductile and brittle states (fig.2.10) tends to show that they are not mutually exclusive. In fact the study of the ductile-brittle transition that has been going on in the last few years has been very rewarding both in the fields of new concepts and in that of practical applications. It rests upon an analysis of the theoretical and microstructural factors that initiate fracturing and propagation of cracks, briefly discussed at the end of this chapter.

Here the macroscopic aspects of fracturing are dealt with in greater detail and we shall look at the relationships between the directions of fracturing and the principal directions of the stress ellipsoid.

3.2. FRACTURE MODES AND RELATIONSHIP TO STRESS

3.2.1. Experimental deformation

The manner of fracturing varies in an isotropic material according to the amount of stress applied perpendicularly to the axis of a test cylinder (fig.3.1). Under atmospheric pressure ($\sigma n = 1 bar = 0.1$ MPa), a sample yields in compression ($\sigma 1 > \sigma n$) by fractures that are largely parallel to the cylinder axis, that is to say in the direction of maximum stress $\sigma 1$. This corresponds to a splitting or parting perpendicular to the plane of fracture, taking place through **extension fractures or cracks** (fig.3.1a). In triaxial tests (fig.2.12b), if the confining pressure σn and the stress $\sigma 1$ ($\sigma 1 > \sigma n$) are simultaneously applied, the fracture is

inclined to the axis of the cylinder and the displacement between
the two sides tends to be parallel to the fracture surface : it
is a shear fracture (fig.3.1b). Under higher confining pressures,
such fractures multiply and their individual shear displacement
diminishes. They become symmetrically inclined to the axis of the
cylinder (conjugate fractures) so that the angle they make with
the axis of the cylinder tends to increase · towards a maximum
value of 45° (fig.3.1c).Finally, at very high confining pressure
deformation becomes penetrative at our scale of observation ; in
a specimen undergoing relatively large deformation, the macroscopic behaviour can be regarded as ductile, although at a microscopic scale the deformation is produced by discontinuous movements
(fig.3.1d). These differing types of behaviour are shown on the
stress/strain curve in figure 3.2 : under confining pressure the
elastic limit is raised whilst the ductile field is extended

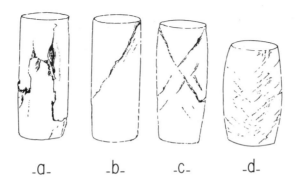

Fig.3.1. Experimental fracturing of marble under increasing confining σn pressures. a) Tension fractures under atmospheric pressure (σ n=0.1 MPa). b) Single shear fracture when σn = 3.5MPa.
c) and d) Conjugate shear fractures when σn =100 MPa (Paterson,
1958. Bull. Geol. Soc. Am., 69, 465).

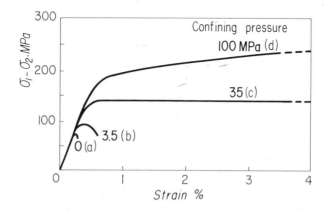

Fig. 3.2. Record of differential stress versus strain, corresponding to tests in fig.3.1. It shows that with increasing confining
pressures, the elastic limit and field of ductile strain both
increase.

before further rupture. In the rock mechanics literature, fractures created by extension are said to follow **mode 1** ; those in
simple shear, **mode 2** and those, in torsion, **mode 3**.

These results can be readily generalised in the case where
the stress ellipsoid has three distinct axes $\sigma 1 \geqslant \sigma 2 \geqslant \sigma 3$: the
extension cracks form in the plane containing $\sigma 1$ and $\sigma 2$, that is
they are perpendicular to the $\sigma 3$ direction, whereas the shear
fracture planes contain the $\sigma 2$ direction and are inclined to $\sigma 1$
and $\sigma 3$ (fig.3.3).

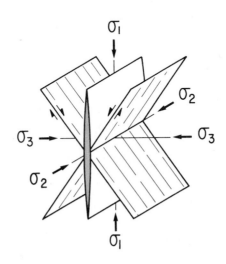

Fig.3.3. Relationship between fractures and normal principal
stresses ($\sigma 1 \geqslant \sigma 2 \geqslant \sigma 3$). Decorated lens : tension fracture ;
striated surfaces : shear fractures.

3.2.2. The Mohr envelope and Coulomb's criterion

It is possible to calculate the shear stress τc which
corresponds to the breaking point of a material under increasing
confining pressure by means of diagrams such as figure 3.2 and by
using the Mohr diagram (fig.2.16). The following analysis is
strictly valid for an **isotropic material**, for example one without
a cleavage which could act as a privileged plane for fracturing.
An experimental rupture curve can be defined as that which measures the maximum resistance τc of the material under different
confining pressures. This line is called the **Mohr envelope** which
represents the maximum shear stresses supported by the material
at the moment of rupture (fig.3.4). The coordinates of the point
of contact allow one to know the maximum normal stress σc and
shear stress τc which can act under given conditions of confining
pressure upon the plane of fracture. The orientation of this
plane, the α angle to $\sigma 1$, can also be found from the diagram.

Under moderate confining pressures and within the field of
shear fracturing, this envelope defines a straight line given by
Coulomb's Criterion : $\tau c = C + \mu \sigma c$. This states that the
resistance of the material in shear is equal to the sum of a
constant C, which expresses the **cohesion** between the crystals,
and of a product in which σc is the normal stress exerted at the

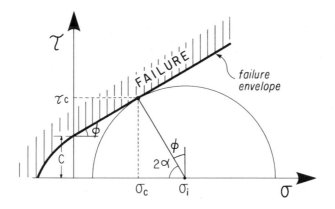

Fig.3.4. Critical conditions for rupture shown on the Mohr's diagram.

moment of rupture and μ represents the **coefficient of internal friction**. Once fracturing has started, displacement on the shear plane will depend upon the roughness of the surfaces which is defined by the coefficient μ . In figure 3.4 it can be seen that C is a measure of the maximum shear stress that the material will support at atmospheric pressure without fracturing and that $\mu =\tan \phi$ defines the slope of the curve of rupture. ϕ is called the **angle of internal friction**, it is also the angle complementary to 2α on the Mohr diagram (fig.3.4). When ϕ is zero, $\alpha =45°$ which is the maximum value and also corresponds to the maximum shear stress upon the plane of rupture (see § 2.3.2, eq.(2)). For example in a diabase deformed under atmospheric pressure, the constants C and μ are of the order of 120 MPa and 1.

Coulomb's Criterion does not apply to the field of **extensional fracture** ($\tau c<C$), that is to say, at very low confining pressures. Equally it does not apply at high confining pressures (fig.3.5). In effect, at high confining pressures the material becomes more ductile (fig.3.2) ; the coefficient of internal

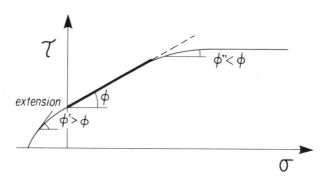

Fig.3.5. Field where Coulomb's Criterion applies (thick line) on the Mohr's diagram.

friction and in consequence the slope of the Mohr envelope diminish. The relationship between τc and σc is no longer expressed
by the linear Coulomb's Criterion equation.
     Under natural conditions, the increase in confining pressure
with increasing depth is accompanied by an increase in temperature (under the action of the geothermal gradient) which also
helps to make the material more ductile. The thermal effect also
diminishes the slope of the Mohr envelope.

3.2.3. Fluid partial pressure and effective pressure

     Consider a porous rock which is permeable and saturated with
water. It is situated at a moderate depth z, let us say at a few
kilometres, where the hydrostatic pressure exerted by the fluid
is Pf~ρh gz, withρ h~1. In this situation, the normal stresses
exerted over any surface area are diminished by Pf with respect
to the value of σn defined in (1) of chapter 2 :

$$\sigma ef = \sigma n - Pf = \sigma i - 1/2(\sigma 1 - \sigma 3) \cos 2\alpha - \rho hgz$$

     σef is the **effective pressure.**

     In this example, we have imagined a porous and permeable
medium in communication with the surface, but this last condition
has little chance of applying beyond a certain depth. From there
on any phenomenon such as compaction, metamorphic reactions and
partial melting will increase the fluid pressure (fig.3.6). This
pressure will rise to that of solid pressure and may exceed it
temporarily if there is an accompanying increase in volume
(§3.2.5).

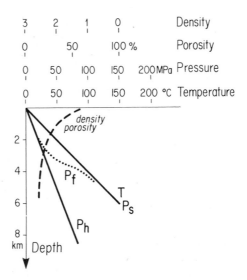

Fig.3.6. Increase in partial pressure of fluids Pf linked to the
loss of permeability during the compaction of sediments (under
conditions of arbitrary P,T). Ps= solid pressure = ρsgz ;Ph
=pressure of the fluid column under consideration = ρhgz
(ρs $\simeq$ 2.5 ρh).

3.2.4. Fracturing assisted by fluid pressure

If a fluid pressure Pf is created in the medium, the centre of the Mohr circle, representing the state of stress in the absence of fluid pressure, is moved towards the left by an amount Pf along the axis of normal stress (fig.3.7). As a result of this translation, the circle representing the state of stress in the presence of fluid can touch the curve of the Mohr envelope, and rupture of the material by fluid assisted fracturing can be predicted. This is a purely physical effect of the fluid, independent of its chemical effects (§3.3.2).

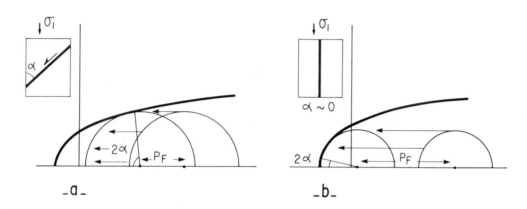

Fig.3.7. Effect of fluid partial pressure Pf on fracturing. a) High stress (circle of large radius) and weak Pf : rupture by shearing. b) Low stress (circle of small radius) and large Pf : rupture in tension.

If in the initial state, the deviator (the radius of the circle of the Mohr diagram) is significant, a small rise in the fluid pressure is enough to cause rupture, moving the Mohr circle into contact with the envelope at a point where the slope is low (fig.3.7a). The angle α is then large and fracturing takes place by shear fracturing. On the contrary if the initial deviator is small, the fluid pressure must approach that of the solid pressure (confining pressure) to cause fracturing ; the movement of the representative circle is important. It touches the envelope near its origin in a zone of steep gradient. The angle α is small and fracturing is formed by extension (fig.3.7b).

Temporarily, a conduit may be opened by fracturing, allowing the expulsion of fluid. In the porous medium, the fluid pressure Pf falls and the fracture closes. Cyclic release of fluid may take place by repeated opening and closing of the fracture like a valve.

Thus porous rocks, where dehydration reactions or partial melting produce a fluid pressure near that of the solid pressure, may yield by assisted fracturing **whatever the depth may be.** When fluids intervene the realm of fracturing is not restricted to shallow depths.

3.2.5. Hydraulic fracturing

When a fluid pressure acts upon impermeable rocks they may break if it exceeds their resistance (fig.3.8). Such a situation may arise when the impermeable rocks surround or overlay a medium in which the fluid pressure exceeds the solid pressure. Among the causes which can produce such overpressures are some dehydration or partial melting reactions involving volume increases (fig.3.8a), or gravitational forces created at the top of a column of a fluid less dense than the surrounding solid medium (fig.3.8b).

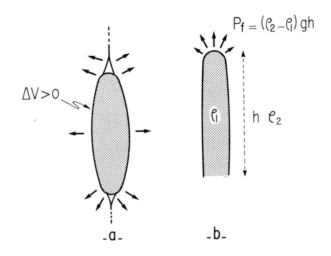

$$P_f = (\rho_2 - \rho_1) gh$$

Fig.3.8. Principle of hydraulic fracturing. An over-pressure in fluid helps to cause fracturing in an impermeable medium. a) Overpressure caused by an increase in volume in conjunction with dehydration or melting reactions taking place in the shaded area. b) Overpressure caused by a fluid column ($\rho 1 < \rho 2$).

An impermeable layer can also impede the flow of fluid towards the surface causing the fluid pressure to reach or exceed the solid pressure. The material underneath may become disaggregated by fracturing and if the fluid fraction exceeds 35 % may turn into a suspension of solids in a fluid matrix. The material is then under-compacted. Such a medium loses its mechanical resistance with its viscosity approaching that of a fluid. Such a liquefied medium can act as a decollement for nappes. After the fluids have been expelled during the nappe movement, the decollement zone consolidates as a tectonic breccia, as in the Cargneulles Formation of the alpine nappes.

Hydraulic fracturing is used in drilling to fracture rocks in situ by injecting a fluid at a higher pressure than that of the rocks' resistance. This can be used profitably to generate or improve permeability locally in the recovery of oil and in geothermal projects, to create or improve the rate of heat exchange between the fluid and the rocks.

3.3. THE BRITTLE-DUCTILE TRANSITION

3.3.1. The formation of microfractures

Most rocks already possess microcavities and fissures that lie along the grains boundaries. If forces are imposed upon them, the stresses are concentrated at the ends of these microcavities from which microfractures or microcracks propagate mainly parallel to the maximum principal stress direction (fig.3.9). Both the number of microfractures, and by coalescence, their size increase in proportion to the stress. As in the case of larger fractures, the angle between the microfractures and the maximun principal stress direction increases with the confining pressure. Rupture occurs when the microcracks join to form a through-going fracture in the material. At low confining pressure, irregular microfractures develop and the rock breaks with a hackly or irregular fracture following the extensional mode 1 (fig.3.1a). At high confining pressure, microfractures are regularly spaced throughout the sample and signs of plastic deformation begin to appear (slip, twinning, kinking, § 4.2.3). Thus rupture at high confining pressure is preceded by a period of slow fracture propagation and of plastic behaviour. Fracturing appears along a zone where the microfractures and traces of plastic deformation are concentrated, following the shear and torsion modes 2 and 3.

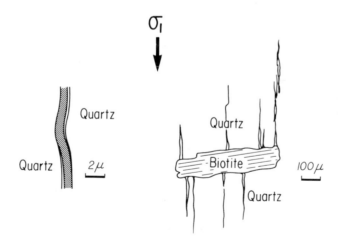

Fig.3.9. Microfractures produced in a granite by the application of a stress $\sigma 1$. a) Old healed fracture which has re-opened. b) Fractures in quartz caused by the deformation of biotite (Tapponnier and Brace, 1976. Int. J. Rock Mech. Min. Sci., 13, 103).

3.3.2. Role of the fluids-stress corrosion

The role played by fluids is not limited to physical effects like creating and keeping fractures open by means of their fluid pressure (§3.2.4 et 3.2.5) or reducing the friction (§2.2.3).

They also have chemical effects, contributing to fracture propagation by corrosion under pressure and to motion along faults by plastic weakening.

The fluid impregnates the microfractures and corrodes or selectively alters the crack tip where stress has built up. This is **stress corrosion**. It promotes crack growth at a lower applied stress than the critical stress required for fracturing (3.2.2) but at a velocity (10^{-3} m/s in quartz) controlled by chemical reaction which is much slower than the rate of catastrophic fracturing (10^{-1} - 10^3 m/s). Water, the most important of the crustal fluids, makes quartz more plastic and assists the formation of alteration minerals (clay minerals, chlorites, micas) which are often ductile. This clay-like alteration along faults is called a **gouge**.

3.3.3. Dilatancy

During axial deformation tests of a rock cylinder with a volume Vo under low temperature conditions and a large deviator , the variations of volume $\Delta V/Vo$ can be measured as a function of increasing differential stress (fig.3.10). A volume reduction takes place first which is due to elastic deformation of the sample, but for larger differential stresses, the volume increases until the sample ruptures. This increase in volume which is called **dilatancy** is due to the opening and multiplication of microfractures (fig.3.1 and 3.9).

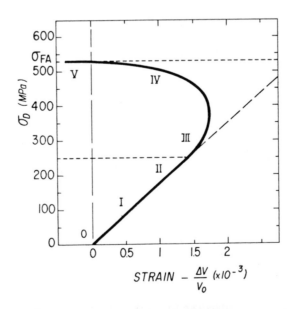

Fig.3.10. Relation between the deviator D and the volumetric strain of a granite under compression. Stages I-II : elastic strain ; stage III : elastic limit ; stages III-IV onset of fracturing ; stages IV-V : dilatancy and rupture.

Confining pressure opposes dilatancy. This is shown by the dissemination of smaller microfractures throughout the specimen resulting in a smaller volume increase. The pattern of fracturing changes in that it is more difficult for microfractures to join up. Also, twinning, kinking and more generally the plastic mechanisms which are activated at high confining pressure do not involve an increase in volume.

As dilatancy precedes rupture it is obviously useful in earthquake prediction. The main techniques under study correspond to the different manifestations of this general effect such as estimation of variations of volume or the increase in stress caused by the dilatancy in the neighbourhood of a fault, the changes in the level and composition of the water table caused by the opening of a network of microfractures modifying permeability and favoring chemical exchanges.

FOR FURTHER READING

Mogi, K., 1973. Rock fracture. Ann. Rev. Earth Planet. Sc., 1, 63-84.

Jaeger, J.C. et Cook, N.G., 1976. Fundamentals of Rock Mechanics. Chapman and Hall, London, 585 p.

Fyfe, W.S., Price, N.J. et Thompson, A.B., 1978. Fluids in the Earth's crust. Elsevier, Amsterdam, 383 p.

Paterson, M.S., 1978. Experimental Rock Deformation - The brittle field. Springer-Verlag, Berlin, 254 p.

never perfect and both crystal growth and deformation have the effect of introducing defects. These defects can be divided into three types : point defects or non-dimensional defects, linear or one-dimensional defects and planar defects or bi-dimensional defects.

Point defects

Point defects are mainly formed by **interstitial atoms** and **vacancies** in the atomic structure. These vacancies play an essential role in the diffusion by exchanging their structural sites with atoms (fig.4.14). The interstitial atoms can be different to those of the structure i.e. impurities, and may be detrimental to the plasticity by causing structural or electrical effects due to differences in volume or electrical charge. However, the presence of water in the lattice of quartz tends to increase plasticity by the hydrolysis of the Si-O bonds in a reaction with hydrogen atoms.

Linear defects

There are domains within crystals where the structure is not continuous with the exterior, often resulting from plastic slip. These domains are surrounded by a **dislocation loop** (fig.4.2) at the boundary between the region that has slipped and that which has not. There are two types of dislocations. The **edge dislocation** results from the insertion of an extra half-plane. The

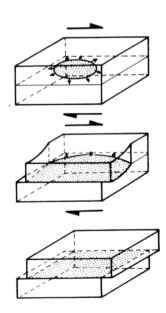

Fig.4.2. Dislocation loop spreading by slip. Under the action of shear stress (arrowed), the slipped surface increases (stippled area) (Nicolas and Poirier, 1976. Wiley-Interscience, London).

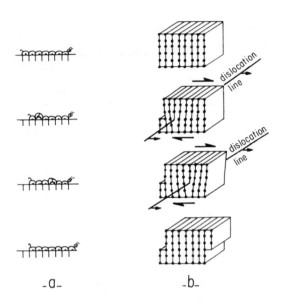

Fig.4.3. Slip by the propagation of an edge dislocation. a) Caterpillar movement. b) Slip in a crystal by movement of an edge dislocation. The extra half-plane corresponds to the ruck in the caterpillar (Nicolas and Poirier, 1976. Wiley-Interscience, London).

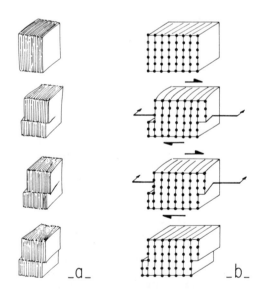

Fig.4.4. Slip caused by the propagation of a screw dislocation. a) Tearing sheets of paper. b) Slip in a crystal by the movement of a screw dislocation. The tear in the sheets of paper corresponds to a screw dislocation (Nicolas and Poirier, 1976. Wiley-Interscience, London).

dislocation is the edge of this half-plane where elastic deformation is greatest (fig.4.3 and 4.5). The **screw dislocation** results
from a helicoidal connection between the reticular planes on one
side and the other of the considered domain and corresponds to
its axis (fig.4.4 and 4.5). The dislocation loop is formed by
edge and screw dislocations of opposite sign which correspond to
half-planes terminating respectively towards the base and the top
for edge dislocations and in spirals of opposite sense for screw
dislocations (fig.4.5). The loop is not usually rectangular ;

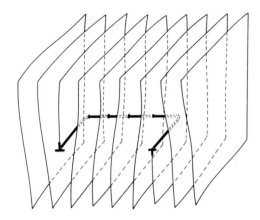

Fig.4.5. Dislocation half-loop. Only the lattice planes parallel
to the extra half-planes are shown. The upper half-plane on the
left is connected continuously to the lower half-plane on the
right by a distortion of the lattice planes bounded at the back
by a screw dislocation, in a similar way to the different levels
in a spiral multi-storey car park. Note the symbol ⊥ used to
indicate the edge dislocations (Nicolas and Poirier, 1976. Wiley-
Interscience, London).

segments with intermediate directions come in-between the pure
edge and screw segments and the loop consequently has segments of
a mixed character. The crossing over of a structure by an edge
dislocation (fig.4.3) or a screw dislocation (fig.4.4) leads to
the shearing of the structure by an amount equal to a reticular
distance in the direction of the screw dislocation. The creation
and spreading of a dislocation loop by dislocation slip within a
crystal are illustrated in figure 4.2. Crystallographically, the
loop plane is called a **slip plane** and the direction of shear,
parallel to the screw dislocation direction, is called the **slip
direction**. The vector displacement associated with slip is called
Burgers vector. It usually corresponds to the reticular distance
of the lattice in the direction of displacement.
    The active **slip systems** in crystals are those which require
the minimum energy to cause the elementary displacement. As
examples, Burgers vectors correspond either to the shortest parameter of the lattice, or to directions parallel to strong chemical bonds, and the slip plane, to a reticular plane of dense
packing or to a plane in which strong bonds need not to be
broken. Olivine (independent SiO_4 tetrahedra) illustrates the

first case with a dominant (010) [100] system corresponding to a dense reticular plane and to the shortest parameter of the lattice ; pyroxene (chains of Si-O parallel to [001] are an example of the second case with a unique system (100) [001] where none of the strong Si-O bonds are broken.The principal slip systems of rock-forming minerals are shown in the table below.

	HIGH TEMPERATURE	LOW TEMPERATURE
olivine	(010) [100]$^+$ (011) [100]$^+$ (001) [100]$^-$	(001) [100]$^+$ (110) [001]$^+$ (100) [001]$^-$ (100) [010]$^-$
kyanite	(100) [001]	(100) [001]
enstatite	(100) [001]	(100) [001] clinoenstatite inversion
diopside	(100) [001]	twinning (100) [001] S = 0.57 (001) [100] S = 0.57
amphiboles	(100) [001]	twinning ($\bar{1}$01) [$\bar{1}$0$\bar{1}$] S = 0.53
micas	(001) [110] (0001) 11$\bar{2}$0 (001) [100]	
plagioclases	(010) [001]$^+$, [100] [101] (001)$\frac{1}{2}$ [110]	albite irrational direction twinning (010) S = 0.14 pericline irrational plane direction [010] S = 0.14
quartz	(10$\bar{1}$0) [1$\bar{2}$10]$^+$ (10$\bar{1}$0) [0001]$^-$ (10$\bar{1}$0) [1$\bar{2}$13]$^-$ (11$\bar{2}$0) [0001]$^-$ (11$\bar{2}$2) [11$\bar{2}$3]$^-$ (1$\bar{1}$01) [11$\bar{2}$0]	(0001) [11$\bar{2}$0]$^+$ Dauphiné and Brazil twinning: polarity changes
calcite	r = (100) [011]$^+$ f = ($\bar{1}$11) [101], [01$\bar{1}$]$^+$ a = (01$\bar{1}$) [011]$^-$	r = (100) [011]$^+$ twinning | e=(011)[100]$^+$ S=0.694 r=(100)[011]$^-$ f=($\bar{1}$11)[211]$^-$
dolomite	twinning f = ($\bar{1}$11) [211]$^+$ S = 0.587	c = (111) [01$\bar{1}$]$^+$ c = (111) [2$\bar{1}\bar{1}$]$^-$
halite	(110) [1$\bar{1}$0]$^+$ (001) [1$\bar{1}$0]$^+$	(110) [1$\bar{1}$0]$^+$ (001) [1$\bar{1}$0]$^-$ (111) [1$\bar{1}$0]$^-$
anhydrite	(001) [010]$^+$	(001) [010]$^+$ (012) [$\bar{1}$21], [1$\bar{2}$1] twinning (101) [$\bar{1}$01]

+ = common   S see fig. 4.9 $^-$ = rare

The elastic deformation of a lattice around a dislocation increases the internal energy of the lattice and creates a local stress field. Two contiguous dislocations with the same sign tend to push apart and two of opposite sign, to attract one another and eventually cancel out (fig.4.13).

As well as being able to slip in their slip plane, the dislocations can also **climb**. In the case of an edge dislocation, climb takes place by the diffusion of atoms located at the edge of the extra half-plane (fig.4.6). If atoms on this plane exchange with vacancies from the lattice, the half-plane is absorbing vacancies and the dislocation line rises. If the vacancies of the half-plane are replaced by atoms its displacement is in the opposite direction. Coordinating these movements by the application of a stress causes deformation.

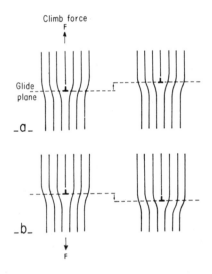

Fig.4.6. Climb of an edge dislocation under the effect of a force F.a) A line of atoms is removed along the edge of the supplementary half-plane. b) A line of atoms is added along this edge (Nicolas and Poirier, 1976. Wiley Interscience, London).

Planar defects

The main planar defects are dislocation walls, twins and the surfaces of the crystal, also called **grain boundaries** when the crystals lying next to each other are formed of the same mineral phase, and **interfaces** when they belong to different phases. Here we shall only look at the internal defects of crystals, namely dislocation walls and twins.

Dislocation walls – The dislocations scattered through the lattice tend to be aligned in a regular way by slip and climb. Such an array corresponding to an energy minimum is a **dislocation wall**. The crystal is thus divided into **subgrains** separated by dislocation walls or **subgrain boundaries** or **subboundaries** (fig.4.7). **Tilt walls** are formed by edge dislocations and **twist walls**, by screw dislocations. The former cause a rotation of subgrains on both sides of the wall defined by a rotation axis contained in the wall and parallel to the dislocation line direction (fig.4.7).

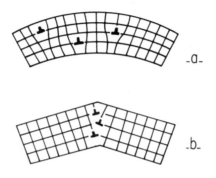

Fig.4.7. Bending in a crystal due to : a) an excess density of edge dislocations with the same sign and b) the location of the same dislocations along a tilt wall (Nicolas and Poirier, 1976. Wiley-Interscience London).

The twist walls are formed of two sets of screw dislocations which cause rotation of the subgrains adjacent to them around an axis at right angle to the wall. Thus, the crystal can accommodate a heterogeneous deformation with rotations of the lattice being confined to the dislocation walls. Such walls are said to be geometrically necessary (fig.4.8).

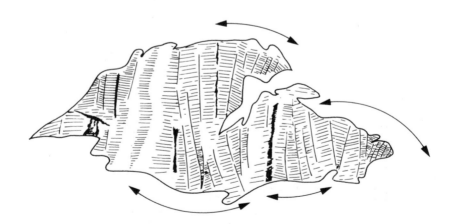

Fig.4.8. Tilt walls in a porphyroclast of olivine deformed heterogeneously (Nicolas et al., 1971. Tectonophysics, 12, 55).

Twins - Twinning affects one part of a crystal giving it a symmetrical orientation with respect to the remainder of the crystal in relation to the twin plane. The twin corresponds to a rotation of a specific amount around a twin axis. In the case of deformation twins, the twinned orientation is derived from the original lattice by a simple shear S of determined angle, parallel to the twin plane K1 and along the direction η 1 (fig.4.9). The lattice in the twinned part undergoes a constant angle of rotation which may be large.

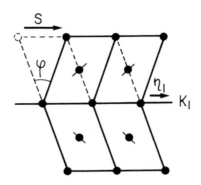

Fig.4.9. Twinning caused by mechanical shearing with amplitude S on the plane K1 in the direction η 1.

4.2.3. Mechanisms and processes of plastic deformation

We shall begin by looking at low temperature mechanisms, that is those operating at temperatures below one-third of the melting temperature. At high temperature they tend to be overtaken by mechanisms and processes of deformation involving diffusion whose activity depends exponentially on temperature.

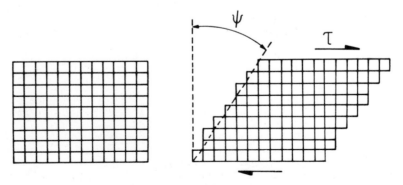

Fig.4.10. Macroscopic shear caused by dislocation slip. τ : shear stress ; ψ :shear angle (Nicolas and Poirier, 1976. Wiley-Interscience, London).

Plastic deformation at low temperature

These are principally gliding, twinning, kinking and cleavage cracking.

Dislocation slip produces macroscopic shear (fig.4.10). In order to deform a crystalline aggregate homogeneously and coherently, each crystal must have five degrees of freedom, that is five independent slip systems (Von Misès criterion). In contrast to the common metals, the principal rock-forming minerals have less than five slip systems due to their lower symmetry. At low tempe- rature, therefore, plastic deformation cannot be homogeneous in the aggregate, slip must be assisted by other mechanisms of deformation such as twinning , kinking and cleavage cracking.

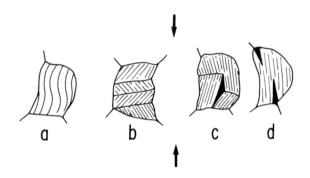

Fig.4.11. Low temperature deformation of crystals having a domi- nant slip direction orientated in the direction of compressive stress (arrowed). a) Deformation by bending. b) Deformation by kinking. c) and d) Deformation by kinking and cleaving parallel to the dominant slip plane.

Compared to slip caused by dislocations, **twinning** takes place very rapidly. It occurs as a local and minor reorganisation of the atoms which requires higher stress conditions than for slip. The rotation angle is fixed and affects the orientation of the lattice whereas in slip, lines attached to the crystal are rotated variably but the lattice is not affected. **Kinking** (fig.4.11b) involves important flexuring of the lattice on each side of the kink which contains the rotation axis. This leads to a shortening of the crystal and its division into **kink bands** which are the domains between two kinks . There is a certain analogy to dislocation walls, although on a microscopic scale the kink plane is more complex. Rotation may also reach several tens of degrees whereas in a dislocation wall it does not usually exceed a few degrees. Lastly, certain crystals cannot deform plastically because they have a particularly unfavourable orien- tation. The concentration of stress may be such that it exceeds the resistance of the lattice. The latter yields along the less strongly bounded crystallographic planes by **cleavage cracking** (fig.4.11c,d and 3.9). It also may fracture completely (brittle fracture). When the confining pressure, which opposes the opening

of cracks, and the temperature rise partial or total recrystallization   allow   these   particular   crystals   to   further   deform
(§ 7.5.2).

Plastic deformation at high temperature

At higher temperatures, diffusion rapidly becomes more active. New mechanisms of deformation such as dislocation climb,
(fig.4.6),   lattice   diffusion   or   surface   diffusion   take   place.
They   increase the freedom of movement (degree of freedom in   the
Von Misès sense) of each crystal,   bringing about a greater homo-
geneity   of   deformation in the rocks. Before looking   at   these
processes in greater detail,   we shall look at the role played by
diffusion in recovery.

Hardening and recovery. At low temperature,   dislocations moving
in   their slip planes tend to be obstructed by various   obstacles
(other dislocations,   impurities,   precipitates and grain boundaries).   For deformation to go on,   stress must be increased ; this
is **hardening** (§2.2.2). This increase in stress causes the generation   of new dislocations in the structure and allows some dislocations   to   overcome the obstacles which held them   up.   On   the
diagram of figure 4.12,   the coefficient of hardening is given by
the slope h = d σ /dε .

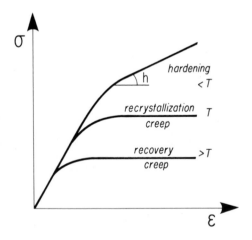

Fig.4.12.   Diagram (σ,ε) illustrating the principal flow   regimes
at different temperatures.

At   high temperatures,   thanks to dynamic   recrystallization
and   **recovery** the congestion of the structure by dislocations   is
reduced.   Recovery   is   the term for the processes   which   reduce
dislocation density,   dislocation interaction and hence   increase
the glide rate. Mutual annihilation occurs when edge dislocations
climb   over   obstacles on one slip plane to meet   and   annihilate

others of opposite sign on another parallel plane (fig.4.13). The
mutual annihilation of screw dislocations can occur at the intersection of co-zonal slip planes or by cross-slip. Unlike the
climb of edge dislocations these interactions are not dependent
on diffusion and hence probably do not control the overall recovery rate. Climb of edge dislocations of same sign into walls
leads to the subdivision of the lattice into subgrains. This is
named **polygonisation by climb** of dislocations.

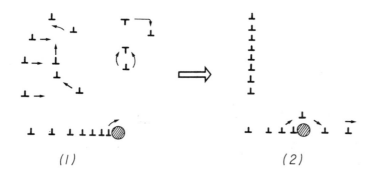

(1)   *(2)*

Fig.4.13. Recovery from 1 to 2 ; above, by annihilation of dislocations of opposite sign and by slip and climb motion into walls;
below, by climb over a dislocation pile-up created by an
obstacle.

Dislocation creep - At high temperature, a dynamic equilibrium
tends to be set up between the opposing phenomenae of hardening
and recovery. The density of dislocations stabilizes which is
shown by a horizontal line on the diagram (σ , ε) (fig.4.12).
This is **steady-state creep**. This type of creep, also called
Weertman creep, is produced by slip and climb of dislocations. It
is said to be controlled by dislocation climb as slip motion
being faster that climb, the flow rate is at most equal to the
rate of climb. In turn the rate of climb is controlled by diffusion. Because of these different factors, the coefficient of
diffusion is introduced into the formula expressing the creep
rate $\dot{\varepsilon}$ (§2.2.2). Thus :

$$\dot{\varepsilon} = A \sigma^n D_V$$

where A = constant ; σ = ($\sigma 1 - \sigma 3$) ; n = 3 ; D_V = diffusion
coefficient depending exponentially on temperature and inversely
on confining pressure.

Diffusion creep. At very high temperature, directed diffusion of
atoms takes place across the crystal lattice or along the grain
boundaries. **Nabarro-Herring creep** implying lattice diffusion acts
under the influence of an applied stress in the way illustrated
in figure 4.14. **Coble creep** implying surface diffusion seems to
occur in the creep of some mylonites with commitment of grain
boundary sliding (fig.4.15 and 8.20). The deformation mechanism
of the aggregate evokes the sliding of sand grains over each

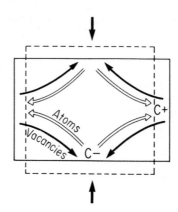

Fig.4.14. Nabarro-Herring flow. A crystal (dashed line) submitted to a vertical compression develops a concentration of vacancies C-, being smaller on faces in compression and C+, being greater along faces in tension. Diffusion of atoms in an opposite direction to vacancies progressively alter the shape of the crystal (solid line).

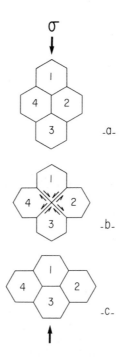

Fig.4.15. Superplastic flow. Gliding along the grain boundaries, as shown by the arrows, aided by surface diffusion, results in a shift of the first neighbours between a) the initial state and c) the final state. The shift is between neighbours 2 and 4 in state a) and 1 and 3 in state c). This results in 55 % strain (Ashby and Verrall, 1973. Acta Met., 21, 149).

other in a bag of sand (fig.4.26). The obstacles are removed by surface diffusion which does not alter the cohesion between crystals. Diffusion is much slower than slip along the grain boundaries and hence it controls the creep rate. This mode of deformation is widely known as **superplastice deformation** in certain alloys and in fine-grained rocks. It is expressed by a creep law strongly dependent upon the grain size d. This law is written as:

$$\dot{\varepsilon} = A'\sigma'Ds/d$$

where A' = constant ; σ =($\sigma'1-\sigma 3$) ; Ds = coefficient of grain boundary diffusion.

The creep rate depends linearly on the stress and thus the viscosity is Newtonian.

Under these conditions and in particular depending on grain size, creep can be controlled by dislocation motion or by grain boundary diffusion (fig.4.16). Thus for a smaller grain size than dc, creep is controlled by diffusion (it is faster by the superplastic mechanism) ; on the contrary for a larger grain it is controlled by dislocation motion.

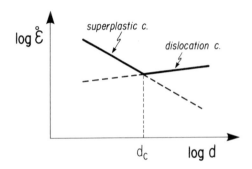

Fig.4.16. Relationship in a diagram (log $\overset{\circ}{\varepsilon}$, log d) between dislocation flow and superplastic flow. For a grain size smaller than dc, deformation is more rapid by superplastic flow ; for a grain size greater than dc, it is faster by dislocation flow.

Recrystallization

Recrystallization can be considered from the view point of its relation with deformation. If it accompanies deformation it is called **syntectonic** or **dynamic** ; if later, it is called post-tectonic or **static** (the **annealing** of metallurgists). It is called **primary static recrystallization** if the energy responsible for the process is the same as in dynamic recrystallization, that is the energy of elastic deformation due to the presence of dislocations. Recrystallization carries on until the dislocations in the recrystallized crystals have almost disappeared.

Another source of energy of recrystallization is the surface energy of the crystals. Their growth, which reduces this energy, leads to **secondary recrystallization**. Because it is driven by reduction in surface energy and not deformational energy, secondary recrystallization can continue after primary recrystalliza-

tion has reduced the dislocation density. Reduction of the surface area of the crystals at the time of the secondary recrystallization firstly takes place by a smoothing of the surface irre-
gularities of the grains, then by the preferential growth of some
grains at the expense of others. If the mobility of grain boundaries is large enough, secondary recrystallization can generate a
mosaic structure in which planar or curviplanar grain boundaries
converge towards triple junctions at 120° (fig.4.17).

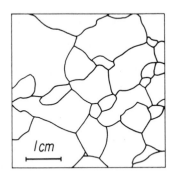

Fig.4.17. Secondary recrystallization structures in ice. Reduction of surface energy leads to the growth of larger grains where
boundaries become curved and converge at 120 to form triple
junctions.

Recrystallization driven by deformational energy operates by
two mechanisms which give rise to recrystallization by **rotation
of subgrains,** and to recrystallization by **nucleation-migration of
grain boundaries.** In the recrystallization by rotation of subgrains, the recrystallized grains (**neoblasts**) are previous subgrains and the grain boundaries, at least in part, former
subboundaries. We have seen how a build-up of dislocations in a
subboundary causes a rotation of one part of the lattice in
relation to the other. Above a certain density of dislocations in

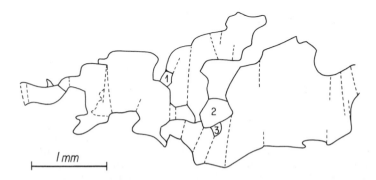

Fig.4.18. Recrystallization of an olivine porphyroclast by subgrain rotation. The neoblasts 1,2,3 have a lattice orientation
still close to that of the adjacent subgrains (Poirier and
Nicolas, 1975. Jour. Geol., 83, 707).

the subboundary, adjacent subgrains in the crystal have rotated up to around ten degrees and the subboundary tends to become a grain boundary. The grain boundary mobility usually does not obscure totally the original structure (fig.4.18). The relation-ship between the deformed host-crystals (**porphyroclast**) and the neoblasts also explains the lattice preferred orientation in the recrystallized structure reflecting the initial preferred orien-tation.

In recrystallization by nucleation and migration of grain boundaries, the neoblasts within the porphyroclasts develop from nuclei by grain boundary migration. At low temperature and high stress, nucleation is favoured over migration due to the intense deformation of the crystalline lattice : on the contrary, at high temperature migration of grain boundaries is dominant. In the first case, the porphyroclasts will be replaced by numerous neoblasts of small size, chiefly localized in the parts of the lattice where deformation has been particularly intense : inside deformation bands or at the periphery of the porphyroclast, forming a mantle (fig.4.19). In the second case, the neoblasts are bigger and less numerous. When grain boundary migration is very active, it directly affects the porphyroclasts : depending on their internal deformation which is a consequence of their orientation, some are growing and others being consumed (§7.5.2).

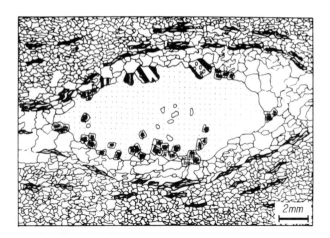

Fig.4.19. Recrystallization by nucleation and growth around a feldspar porphyroclast. Little neoblasts develop in bands and grow preferentially at the periphery of the porphyroclasts (Debat, 1974. Thesis Toulouse).

As in the case of recrystallization by rotation of sub-grains, recrystallization by nucleation-migration leads to preferred orientations inherited from the previous state, but usually more diffuse, especially at low temperature.

Figure 4.20 gives estimates of the temperature and stress conditions at which different styles of recrystallization are active. At high temperature, recrystallization by nucleation-migration may be dominant with respect to recovery and recrystal-lization by rotation of subgrains if the applied stress is high ;

the deformation rate then being very high, recrystallization becomes a more effective process than polygonization for reducing the density of dislocations ; an opposite situation to that, prevailing at low stress.

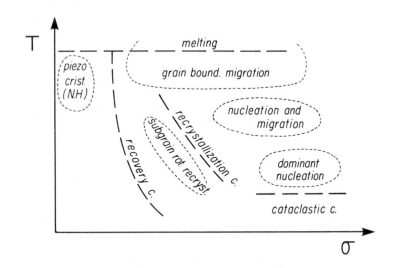

Fig.4.20. Diagram (T,σ) showing the fields of the principal mechanisms of recrystallization and their relationships to the different types of creep (c).

Structural piezometers

We have mentioned in § 4.2.2 that because of elastic deformation in their vicinity, two dislocations with a similar sign repel each other. An internal stress controls their spacing, measured by the density of dislocations in the crystal. Assuming that this internal stress may be equated with the deviator σ applied upon the crystal leads to the relation : $\sigma = K1\,\mu\,b\sqrt{\rho}$, where : K1 is an experimental constant ; μ is the rigidity modulus ; b is the Burgers vector ; ρ is the dislocation density.
One can establish another relationship between the stress and the average spacing d of the subboundaries : $\sigma = K2\,\mu\,b/d$ where K2 is a new experimental constant.
A similar relation can be found between the stress and the average size d' of the neoblasts, resulting from subgrains rotation since the neoblasts form directly from the most disoriented subgrains.
Lastly, the previous paragraph accounts for the well-known relationship in metallurgy between the stress and the size dn of the neoblasts resulting from recrystallization by nucleation-migration of grain boundaries : $\sigma = K3\,\mu\,b/dn$, where K3 is a new experimental constant.
These piezometers are obviously of interest. Thus in the creep law : $\dot{\varepsilon} = A\sigma^{n}D$, the stress may be the only unknown parameter, if the others, A, n, as well as those which come under the definition of the coefficient of diffusion D, have been

measured experimentally and if the T and P conditions of the deformation under consideration are known by geothermometric and geobarometric methods. For example an estimation of σ allows the calculation of $\dot{\epsilon}$ and the relative displacement velocity at the boundaries of a mass deforming in simple shear if its thickness is also known.

By way of illustration we can analyse the case of a shear zone in ophiolitic peridotites which is assumed to represent a transform fault in the oceanic lithosphere where the ophiolite originated. The zone has vertical foliations over a total width of 10 km and horizontal lineations due to the sliding of one plate against the other. The temperature and the pressure of equilibration, deduced from studies of the pyroxenes, are of the order of 1000 C and 300 MPa respectively. The piezometers indicate a normal deviatoric stress of the order of 100 MPa, which corresponds to a shear stress $\tau = 50$ MPa ($\S 2.3.2$).

Introducing these data into the creep law which has been experimentally established by Post it is possible to calculate the shear rate per second :

$$\dot{\gamma} = 1.7.10^9 \tau^{3.2} \quad \exp(1.16 \times 10^5 + 2.63 \ P)/RT$$
(P in MPa ; T in $^\circ$ K ; R = 1.987 cal dg^{-1}).
$$\dot{\gamma} \simeq 3.10^{-14} \ s^{-1}.$$

If it is assumed that the movement in the shear zone is homogeneous it is possible to calculate the velocity V of the sliding of one plate against another :

$$V = \dot{\gamma} \ x10^6 \ cm$$
$$V \simeq 1 cm/year$$

Unfortunately, the structural piezometers are still unreliable. Thus, piezometric measurements based upon the direct observation of free dislocations are biased by the fact that this dislocation substructure is very easily altered by events subsequent to the deformation under consideration (recovery, small deformations under a stress pulse,etc). The most reliable piezometers in this respect rest upon measurements of the size of the neoblasts and of the most disoriented subgrains. But the difficulties of experimental calibration (determination of K) for these piezometers, are still unresolved.

4.3. FLUID ASSISTED DEFORMATION

We shall not speculate upon the nature of the fluid which may be hydrous and magmatic. The presence of a fluid which wets the grain boundaries and interfaces increases surface diffusion considerably. Under an applied stress, the change in form of a particular crystal is more rapidly accomplished by **diffusion** than by dislocation motion. Plastic deformation by dislocation motion no longer takes place or only in an accessory fashion. A fluid presence also favours **fracturing** (Chapter 3). After a rapid examination of the deformation mechanisms by diffusion or fracturing, we shall see how their joint action accounts for aggregate deformation.

4.3.1. Pressure-solution

Suppose a single phase aggregate is placed under a stress ($\sigma 1-\sigma 3$) such that each crystal is locally in contact with its neighbours and moistened by a fluid (fig.4.21). The equilibrium concentration of vacancies is not the same on the perpendicular faces at $\sigma 1$ and $\sigma 3$, thus creating a chemical potential gradient. This is higher at A than B (fig.4.21) and the crystal will tend to dissolve on the A faces and to grow along the B faces by diffusion of atoms within the fluid phase.

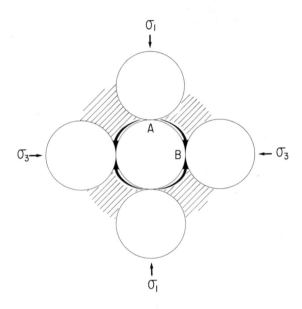

Fig.4.21. Mechanism of deformation by solution-crystallization. Under the deviator ($\sigma 1-\sigma 3$), the central grain dissolves at A and the material is transported by a fluid to B where it is deposited (fluid zone hachured).

As shown in figure 4.22, this mechanism involves three stages : oriented solution, fluid transport and deposition of the dissolved elements at variable distances.

Solution

Within quartzites and limestones evidence for the pressure-solution mechanism is provided in favourable cases where the shapes of the original grains or pebbles can still be seen (fig.4.22 and 4.23). Solution can also be discontinuous or periodic (fig.7.11). Selective dissolution leads to a residual concentration of insoluble minerals which are often dark coloured and stand out as blackish bands and dark fringes. In an anisotropic or previously fractured medium, solution may be more active along surfaces where circulation is easy, and thus more loosely related to $\sigma 1$.

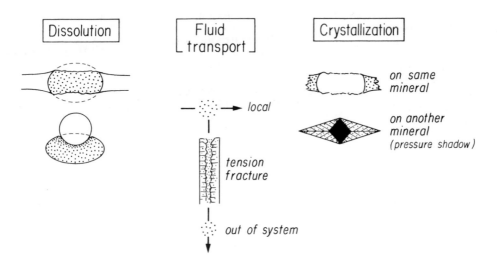

Fig.4.22.   The   three   stages   of   deformation   by   solution-
crystallization (modified from Ramsay, 1980. Carbonates, AGSO,
Bordeaux, 389).

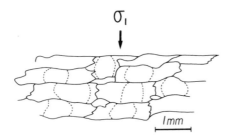

Fig.4.23.   Natural   deformation   of   a   quartzite   by   solution-
crystallization. The former outline of the quartz grains is shown
by lines of impurities (dotted).

Transportation : fluid-assisted fracturing

     Transportation   within   a   fluid can take   place   either   by
diffusion in an immobile fluid,   or by transport within a   moving
fluid in a fracture or by a combination of these two modes.
     Although   hydraulic   and   fluid-assisted fracturing is not   a
mechanism of continuous deformation, its role must also be considered here.   In effect some microstructures that have been attributed   to fluid fracturing (fig.4.24) may lead to   a   penetrative
macroscopic deformation. Furthermore, study of the microfractures
suggests   that   the opening of fractures,   where the material   in
solution   is deposited ,   is controlled by the rate   of   chemical
reactions which implies that deformation is slow and continuous.
     In   chapter 3 we saw that the partial pressure of the   fluid
within   a   porous   medium,   by reducing the   effective   isotropic
pressure, leads to fracturing of the rock. Due to the creation of

a local fluid overpressure, for example during a dehydration
reaction, tensional fracturing can extend through an adjacent
non-porous medium (hydraulic fracturing in the narrow sense). In
a porous rock which is under stress and undergoing dissolution,
the supersaturated fluid can precipitate the dissolved materials
along fractures which it has helped to open. This deposition
evidently contributes to the deformation. Such a mechanism can
open large and spaced fractures (fig.5.9), or at a microscopic
scale, open progressively innumerable small parallel cracks
between the grains. Once opened, these microfractures tend to
seal due to the growth of minerals across them, this growth being
active in proportion of the oversaturation of solutions in
dissolved elements. With the sealing of the fractures, the fluid
pressure increases once again until a new microfracture forms.
This is the **crack-seal** process. The cycle goes on until the fluid
is exhausted. The result is a macroscopic deformation and the
appearance of a cleavage due to the oriented growth in the cracks
of new minerals (fig.4.24).

Fig.4.24. Development and deformation of a cleavage by the crack-
seal mechanism in a medium rich in saturated fluids. The growth
of minerals at high angles to the walls of the microfractures
accentuates the appearance of the cleavage (S).

Crystallization

The elements released by the dissolution of a mineral can
contribute to the deposition of an identical or different mineral. In the first case, the crystallization of the dissolved
mineral may precipitate elsewhere in crystallographic continuity
on the same crystal (fig.4.21 and 4.22), on another crystal of
the same phase following a different crystallographic orientation
or on a grain of another phase. The crystallization is often
fibrous in habit with elongation in the $\sigma 3$ stress direction,
this is deposition in **sheltered zones** and **pressure shadows**

(fig.4.22, 8.3 to 8.5). Lastly, crystallization can take place along extension cracks at variable distances from the site of solution (fig.4.22) in relation with the process of fluid fracturing.

4.3.2. Passive rotation

A rock is often composed of minerals that are soluble and others that are insoluble in the presence of the fluid under consideration. Insoluble minerals of average grain size in the presence of an interstitial fluid are not affected by the solution-crystallization process and may undergo a **passive rotation** (fig.4.25). This may be accentuated when there is a loss of volume during the deformation, as in the case of compaction or where the dissolved material is not precipitated in situ. Thus within pelites and marls, dissolution preferentially affects quartz and calcite whilst the micaceous minerals are reoriented passively. This passive rotation does not adequately explain the very regular cleavage of slates and phyllites. Micaceous minerals also **crystallize** or recrystallize directly in the cleavage plane, their large growth anisotropy helping to reinforce the preferential orientation.

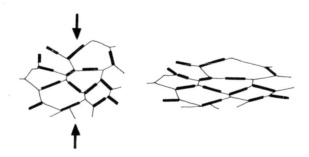

Fig.4.25. Cleavage development by a passive rotation of planar minerals. This rotation is induced by solution-crystallization affecting the matrix.

4.3.3. Grain boundary sliding and liquefaction

A fine-grained aggregate immersed in a fluid under pressure (Pf≈Ps) can also deform by sliding along the grain boundaries accomodated by surface diffusion (fig.4.26). The nature of the medium in which diffusion takes place (intergranular fluid rather than a solid medium) distinguishes this mechanism from that of superplastic deformation affecting solid media (§ 4.2.3). An extreme case akin to intergranular sliding is that of deformation by **liquefaction** (§ 3.2.5).

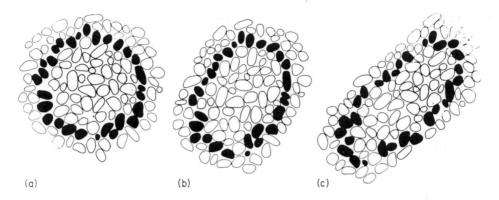

Fig.4.26. Deformation by grain sliding from a) to c), either
assisted by diffusion in a fluid medium if the medium is porous
but compact, or solely by grain sliding if the medium is
undercompacted (Borradaile, 1981. Tectonophysics, 72, 305).

FOR FURTHER READING

Durney, D.W., 1976. Pressure-solution and crystallization deformation.
Phil. Trans. R. Soc. Lond., 283, 229-240.

Nicolas, A. et Poirier, J.P., 1976. Crystalline plasticity and solid
state flow in metamorphic rocks. Wiley - Interscience, London, 444 p.

Poirier, J.P., 1976. Plasticité à haute température des solides
cristallins. Eyrolles éd., Paris, 320 p.

Poirier, J.P., 1985. Creep of crystals. Cambridge University Press,
Cambridge, 260 pp.

Ramsay, J.G., 1980. The crack-seal mechanism of rock deformation. Nature,
284, 135-139.

Chapter 5

Discontinuous Deformation: Structures, Interpretations

5.1. INTRODUCTION

The brittle and ductile behaviours correspond respectively to discontinuous deformation and continuous deformation (§2.1.1, 2.2.2. and fig.2.2). Natural manifestations of discontinuous deformation include faults and fractures in which the nature of the displacement is different. First we shall consider those discontinuous structures caused by no or little displacement parallel to the surface : joints, tension fractures, stylolitic joints (which are not due to brittle deformation but are associated with the preceding structures), then the faults which correspond to large displacements along their surface. The associations made by these different structures and their relations with the states of stress will also be considered.

Faults, ductile faults and shear zones (§8.3) form a continuum. According to the competence of the medium involved, behaviour can be purely ductile, ductile-cataclastic (see §4.2.1) or ductile-plastic sensu-stricto. At a crustal scale, deformation by faulting is dominant in the superficial formations, passing with depth into continuous deformation (fig.5.16). Studies of rheology and of deformation in the brittle-ductile transition field (§2.2.2) help us to understand such behaviour in the field. This chapter is related with chapter 8 on shear zones.

Lastly we have emphasized in chapter 3 the numerous practical applications of the study of discontinuous deformation. We shall show briefly how tension fractures generate mineral veins and explain the circulation of magmas through the conduits that feed volcanoes. The relation between faulting and seismic activity will also be raised. **Seismotectonics**, a new branch of tectonics specifically deals with this problem.

5.2. JOINTS

Joints are planes of parting in rocks, which involve neither displacement nor filling (dry fractures). Joints which are commonly parallel and perpendicular to the bedding plane form a network that split the rock up into large prisms (fig.5.1). They form in competent rocks such as limestones, sandstones, or eruptive rocks. Fracture cleavage can be related to jointing (§6.2.3). These discontinuity surfaces can have different origins. The prismatic parting, planar or conchoidal, found in volcanic rocks is attributed to contraction of the rock during cooling. Networks of joint as well as fracture cleavages are particularly well developed in faulted and folded zones, showing

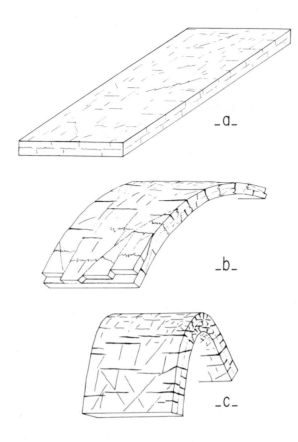

Fig.5 1. Formation of joints in a limestone bed under shallow
cover. a) Network of joints in an undeformed limestone. b) Formation of joints parallel to the fold axis either as tension fractures (black),   at the outer edge of the bed, or as stylolitic
joints within it ; the oblique joints locally evolve into slip
surfaces (faults). c) Amplification of the preceding structures ;
the stylolitic joints are concentrated on the inside of the arch;
elsewhere joints and tension fractures (Droxler and Schaer, 1979.
Eclog. Geol. Helv., 72, 551).

that   there   is   a relationship between their formation and   the
stress   pattern   (brittle   relaxation   of   elastic   deformation)
(fig.5.1). Finally, in massive and competent rocks such as gra-
nites, a network of joints form parallel to and near to the
surface, also called **exfoliation** ; it is attributed to relaxation
of the isotropic or mean stress (fig.5.2). In effect, when a
granite is progressively exhumed by erosion, the vertical stress
component drops until it reaches atmospheric pressure (O.1 MPa)
and elastic deformation of the granite, due to release of stress
in a vertical direction, causes exfoliation.
     As shown in fig.5.1, some of the joints that appear early in
rocks can later be sites of solution, tension fracturing or faulting.

Fig.5.2. Exfoliation in an outcrop of granodiorite. Note the
increase in jointing towards the surface and their independence
of the steeply dipping magmatic lamination.

5.3. TENSION FRACTURES

Tension   fractures,   fissures,   dikes   and   veins   are
distinguished from joints by an infilling which is evidence of a
dilation (fig.5.1,   6.7,   8.23 and 8.24).   Analysis of the fiber
structure of infilling minerals (§ 8.2.3) and of the comb   structure found in magmatic veins indicates that displacement is often
perpendicular to the direction of the fracture. A component of
displacement by shear,   parallel to the plane of the fracture can
also appear (oblique tension fractures, fig.8.10). This component
becomes   dominant in faulting where however tension can intervene
locally.
In   the field,   the tension fractures form a system of fissures   and   tapering   parallel   veins   often   lying   en   echelon
(fig.5.3).   En echelon veins may also occur between two parallel
faults or shear bands forming a corridor.   The angle between   the
faults   or   individual fissures and those of the band   is   around
45 , in conformity with the analysis of § 5.5.6. In such a system,
the   lenses tend to be relatively short and wide (fig.5.4).   In a
conjugate   system,   the faults and veins are organized   into   two
directions   symmetrically inclined to the main deformation directions (fig.5.5).
If the brittle fracturing is accompanied by some   ductility,
the   en   echelon fissures may deform,   recording the   history   of

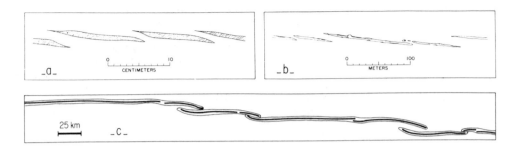

Fig.5.3. En échelon tension fractures at various scales ; a) and
b) are **left-stepped** fractures and c is **right stepped**. a) Quartz
veins. b) Minette dikes. c) Map of the East Pacific Rise ; solid
line, zone of volcanic emissions ; elsewhere, submarine plateau.
(a) and b) Pollard et al., 1982. Geol. Soc. Amer. Bull., 93,
1291.; c) Lonsdale, 1984. Geol. Soc. Amer. Bull., 96, 313).

deformation. The pattern observed depends upon the ways the
fractures have opened and the exact location of the ductile
shearing (fig.5.6).
      The tension fractures may terminate in joints and hair line
fractures (fig.5.7). They are often associated with faults and
shear zones. The attenuation of the relative displacement in
individual faults and shear zones at the tips of the dislocated
area can be achieved in several ways : by penetrative plastic
deformation (fig.8.15), by smaller shear zones (fig.8.16) or by
pressure solution figures on the compressive side, and by tension
fractures on the extensive side ; thus in the case of figure
5.8a, a single and wide tension fracture is associated with
stylolitic joints on the compressive termination. The motion on
such a fault is obviously slow, the deformation rate being controlled by pressure-solution. In contrast, tension fractures may

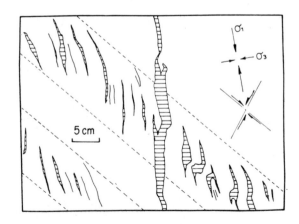

Fig.5.4.   En échelon and lenticular quartz fractures, in
quartzite. Dashed lines indicate limits of shear bands (Roering,
1968. Tectonophysics, 5, 107).

form a horse-tail pattern of thin, long and often dry fractures
(fig.5.8c and d) favouring coalescence between individual faults
(fig.5.8c). Both the longitudinal (direction of fault motion) and
the transversal terminations of a fault can be achieved by tension fractures which thus surround the dislocated area
(fig.5.8b).

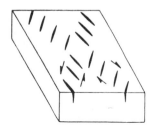

Fig.5.5. Conjugate system of en échelon lenses.

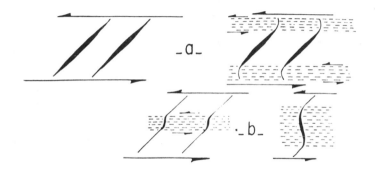

Fig.5.6. Rotation of tension lenses by ductile shearing, a)
confined to the margins of the band, b) localised in the centre
of the band (Roering, 1968. Tectonophysics, 5, 107).

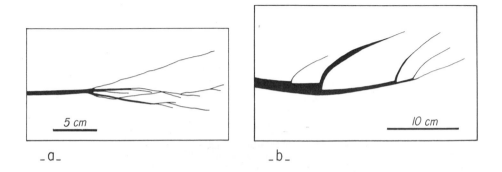

Fig.5.7. Ends of tension fractures, a) symmetrical, b) asymmetrical (Beach, 1980. Jour. Struct. Geol., 2, 425).

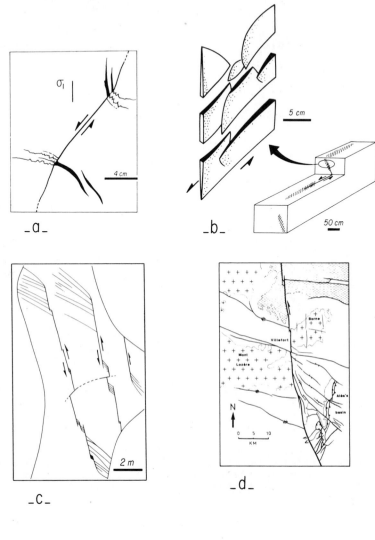

_a_

_b_

_c_

_d_

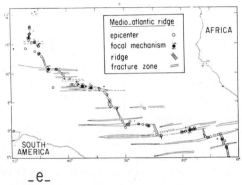

_e_

Study of the dynamics of tension fractures shows that they form in the plane of the principal stress directions ($\sigma 1$, $\sigma 2$) (§3.2 and fig.3.3). Where the study of fibrous minerals (§8.2.3) indicates that a component of shear was accompanying the opening, the Mohr analysis (§2.3.2 and §3.2.2) predicts that the plane of fracturing becomes oblique at an angle α to the plane of ($\sigma 1$, $\sigma 2$) (fig.2.15). For an angle $\alpha < 45° - \phi/2$ (fig.3.4 and 3.5), a tensile normal stress can affect the plane of fracture because the centre of the Mohr circle can have a negative abscissa. For $\alpha \geqslant 45° - \phi/2$, the normal stress becomes compressive. This angular limitation depending upon the ϕ value of the material (about 22° for dolerite) marks the transition from tension fracturing to shear faulting where compressive and tangential movements are dominant.

Another experimental finding brought to light by the Mohr analysis is that, in the absence of fluids which are able to exert a partial pressure, tension fractures can only form near to the surface ($\sigma i \simeq 0$ on fig.3.4). On the contrary, the presence of fluids can cause rupture at whatever depth (§3.2.4). Therefore we accept that tension fractures are, with the exception of the most superficial ones, produced only by assisted or hydraulic fracturing. This conclusion is supported by the synkinematic (that is to say synchronous with movements) and non-secondary character of mineral infilling in these fractures (§8.2.3). In metallogeny, hydraulic fracturing thus controls vein deposits, which in structural terms are infillings of tension fractures.

5.4. STYLOLITIC JOINTS

Stylolitic joints are rough surfaces finely covered with little peaks, the stylolites which are outlined by a concentration of phyllitic and opaque minerals (fig.5.9). These joints are particularly common in limestones where they are formed by a concentration of residual mineral matter following the pressure-solution of a particular horizon (§4.3.1) (fig.5.10). The peaks probably result from varying rate of solution, which itself is dependent on a subordinate irregular distribution of insoluble impurities. They are thus parallel to the direction of shortening.

Fig.5.8. Relations between tension fractures and faults. a) Termination of a small fault in limestone giving stylolitic joints on the compressive side and calcite-filled tension fractures (black) on the extensive side. Their orientations with respect to the expected orientation of the external maximum stress are explained by local stress deviations (see fig.5.22). b) Terminations of a fault in en échelon tension fractures with detail of the transition between the shear and the tension surfaces. c) and d) Respectively small and large scale horse-tail fractures in a granite ; note in c) the coalescence of faults and in d) the presence of Alès Basin on the dilational segment of the Villefort fault which may result from the dynamics of this system. e) Relationship between an ocean ridge and transform faults. (a) Rispoli, 1981. Thesis Montpellier ; b, c and d) Granier, 1985. Tectonics, 4, 721 ; e) Sykes, 1967. Jour. Geophys. Res., 72, 2137).

Because of their mode of formation, the stylolitic joints tend to be orientated perpendicular to the stress $\sigma 1$, with the stylolites pointing in that direction. The stylolitic joints are thus frequently perpendicular to the extension fractures (fig.5.9). It is not always so, particularly in the case of stylolites developed along faults, where they can point in the direction of displacement, contributing to the striations along the fault-planes (fig.5.17c).

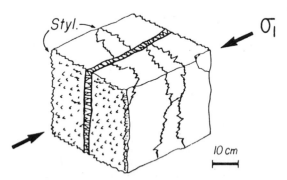

Fig.5.9. Stylolitic joints and fractures infilled with calcite in a limestone showing the relationship to stress $\sigma 1$.

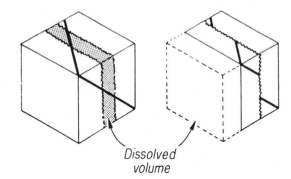

Fig.5.10. Formation of stylolitic joints by selective solution (dotted area). The apparent displacement of the marker gives the amount of shortening.

5.5. FAULTS

Faults are surfaces of discontinuity parallel to which the principal displacement has occurred. The magnitude of the displacement vector or throw of the fault can be determined if a sufficient number of displaced markers are known ; **striations or striae** (§ 5.5.2) grooved upon a smoothed fault surface of easy parting (a **slickenside**) show the direction and sense of movement.

5.5.1. Geometrical analysis of faults and fault systems

The principal types of faults are shown in figure 5.11.
Where the fault plane is inclined, the lower surface is called
the **footwall** and the upper surface the **hanging wall** ; in a **normal
fault**, the hanging wall is raised up (fig.5.11a) and in a **reverse
fault**, it drops down (fig.5.11b). In these faults the vertical
component of the displacement (upthrow) is dominant. On the
contrary in a **strike-slip or wrench** fault (fig.5.11c) the overall
displacement is horizontal. A strike slip fault is **dextral** or
right-lateral if the wall away from the observer has been displaced to the right (or clockwise) , and **sinistral** if it has been
displaced to the left. An inclined normal fault, apart from its
vertical displacement, creates a horizontal extension which is
all the more important when the fault plane has a low dip. Similarly, a reverse fault creates a horizontal shortening.

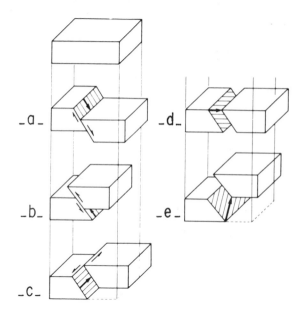

5.11. Principal types of faults. a) Normal fault. b) Reverse
fault. c) Sinistral strike-slip fault. d) and e) Composite
faults, d) sinistral normal fault, e) sinistral reverse fault
(Mattauer, 1973. Hermann éd., Paris).

Sometimes faults form conjugate systems. Thus one can find
systems corresponding to extensional (fig.5.12a), or shortening
(fig.5.12b) or indentation tectonics (fig.5.12c). On a crustal
scale, the rift valleys are extensional systems ; their evolution
to passive margins results from considerable crustal stretching.
Analysis of a rift-passive margin system suggests that the normal
faults become shallower with depth ; the sedimentary beds
initially horizontal rotate to a corresponding angle (fig.5.13).

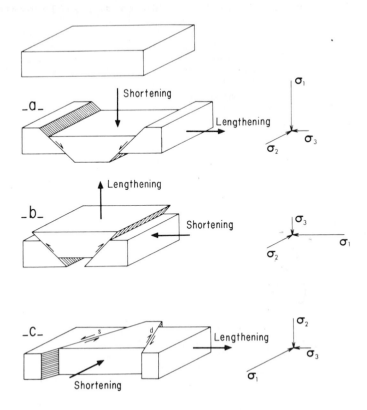

5.12. Systems of conjugate faults. a) Conjugate normal faults
forming a graben. b) Conjugate reverse faults forming a horst. c)
Conjugate dextral and sinistral strike-slip faults with indentation (Blès and Feuga, 1981. BRGM éd., Orlèans).

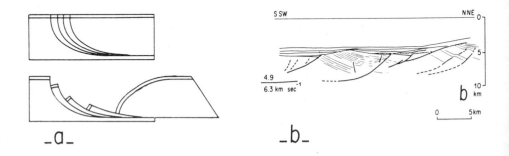

5.13. Rotational faulting of beds caused by the progressive
shallowing of normal faults with depth. a) Model ; note the
bending of the right hand side border of the faulted domain. b)
The passive western Armorican margin ; the sedimentary beds (fine
lines) were detected by seismic reflection. (a) Wernicke and
Burchfield, 1982. Jour. Str. Geol. 4,105 ; (b) Montadert et al.,
1979. Int. Rep. Deep Sea Drill. Proj., XLVIII).

The horizontal component of the displacement becomes progressively greater than the vertical component. Accordingly, from discontinuous on the surface, the deformation seems to pass at depth into a continuous deformation which, on the scale of the crust, corresponds to a regime of horizontal stretching. A comparable transition exists between superficial reverse faults and the deeper ductile thrusts (fig.8.17). Also, in crustal history, shear displacements tend to be repeatedly located along preexisting shear zones (§8.3.2).

Extension   Compression

Fig.5.14. Tensional and compressional sectors along a fault with a curved surface.

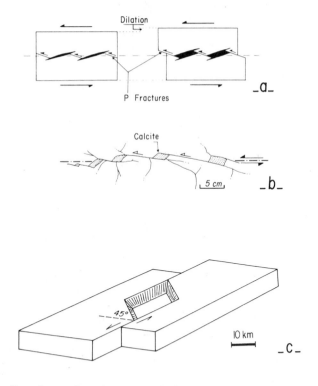

Fig.5.15. Tension fractures and dominoes related to non-planar faults. a) and b) Tensional dominoes formed by displacement on a fault where the surface is made up of alternating facets of shear P and of fracture T. a) Theoretical model. b) Natural fractures. c) Formation of sedimentary basins by the same mechanism operating on a crustal scale along a major fault at the scale of the crust.(a) and b) Gamond, 1983. Jour. Str. Geol., 5, 33).

5.5.2. Internal structures of faults

Whether the fault surface is very smooth or made up of a series of small fractures, it is generally warped. The movement of the two sides thus determines sectors of extension and of compression (fig.5.14). The extension sectors open like a set of lenses or dominoes (fig.5.15a,b,c) aligned parallel to the surface of the fault and, in the case of small openings, filled with similarly aligned fibrous minerals (fig.5.18).

The appearance of the compressional sectors depends upon the way in which obstacles to the movement have been removed (fig.5.16). If deformation is brittle, fracturing and grinding of the rocks result in a **breccia** or **microbreccia**, formed from rock fragments of varying size and more or less cemented by infilling minerals. Commencing at an adequate depth (see § 3.2.3.), it is thought that brecciae result from hydraulic fracturing and that during movement the medium is undercompacted and liquefied (§ 3.2.5). The fluids can also induce continuous deformation by solution of exposed parts of the fault-planes which are subjected to a concentration of stresses. Stylolites are evidence of such solution (fig.5.17c). At depth, in **ductile faults** and in shear zones, plastic deformation takes the place of the preceding mechanisms. This is shown by the development of **mylonites** and **ultramylonites**. Their minute grain size (sub-microscopic in the ultramylonites) is due to cataclasis and dynamic recrystallization (§ 4.2.1 and 8.3.2). Finally, frictional fusion can lead to a local infilling by **pseudotachylites** (§ 8.3.2).

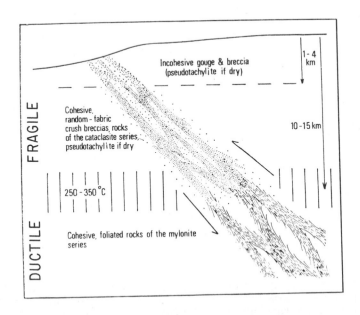

Fig.5.16. Modes of deformation as a function of depth along a major reversed fault (modified from Sibson, 1977. Jour. Geol. Soc., 133, 191).

5.5.3. Seismic and aseismic faults

The ductile deformation structures just described are characteristic of faults moving at geological velocities (a few cm a year). These faults are called **aseismic** or **non-seismic** as opposed to **seismic** faults where the displacement velocity is sudden (in the order of km/s). The structures associated with the latter are of a type corresponding to violent rupture (forming breccias and pseudotachylites). The same fault may show both types of structures, either superimposed upon one another in the same place, implying a succession of episodes of continuous and discontinuous deformation along its own fault-line, or one relaying the other along the fault trace, suggesting that some sections can slide in a continuous way whilst others, normally immobile, yield abruptly and violently. After each episode of rupturing, the deviatoric stress is relaxed. It then builds up progressively before a new rupture can take place. This is called **stick-slip**. The San Andreas fault has both of these types of deformation. In some sections, it slips regularly 1 to 2 cm a year whereas in the neighbourhood of San Francisco where it is locked, the last displacement in the 1906 earthquake was about 5m.

5.5.4. Microstructural analysis of displacement

Structures linked to faulting constitute an imprint of displacement. Striae are used mostly, but also flexuring of the walls (fig.8.18a) and possibly drag folds (fig.8.18c).

Striae indicate the sense of displacement and more precisely, that of the last displacement. They are often asymmetrical, showing facetting and roughness in one direction giving a "nap" to the fault. It can be related to the sense of displacement if the cause of the striations can be identified. It may be due to the fibrous infilling (fig.5.17b), to stylolites (fig.5.17c), to a hard object attached to one side of the fault, making a groove on the other (fig.5.17d) or to second order fractures P or R (§5.5.6 and fig.5.21, 5.23). Infilling fibers and stylolites can face one another as shown in figure 5.18.

5.5.5. Dynamic analysis

In the case of isotropic rocks the general relationship between shearing along a fault and the principal stress directions is known, as a result of experimental (fig.3.1) or natural deformation and is described well by Mohr's analysis (§2.3.2 and 3.2.2). Figure 5.12 shows this relationship for the main types of faults. The principal stress directions can be defined if measurements can be made upon two conjugate faults. The direction of $\sigma 2$ is parallel to the intersection of the two faults and that of $\sigma 1$ is contained within the bisector of the acute angle between the two faults (actually the analysis in §3.2.2. shows that fracturing cannot occur on a plane whose angle with $\sigma 1$ is greater than 45°). The direction of $\sigma 1$ estimated in this way must coincide with that which can be deduced from the shear sense under the criteria given in § 5.5.4. The determination of the $\sigma 1$ and $\sigma 3$

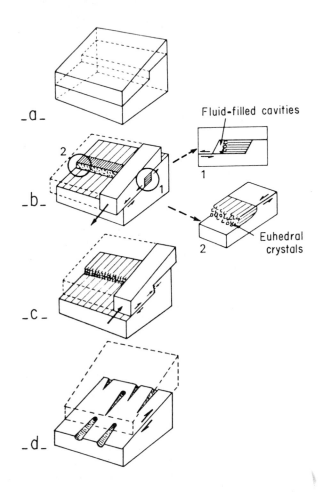

_a_

Fluid-filled cavities

2

_b_

1

1

2

Euhedral
crystals

_c_

_d_

Fig.5.17. Striae giving the sense of movement on a fault plane.
a) Stepped fault-plane.   b) Fibrous minerals related to a
sinistral normal fault. c) Stylolites related to a dextral reverse fault. (modified from Mattauer, 1973. Hermann éd., Paris).

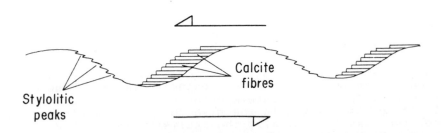

Calcite
fibres

Stylolitic
peaks

Fig.5.18. Fibrous calcite crystals fibres and stylolites peaks
facing each other on curved portions of a sinistral fault.

directions is approximate or well nigh impossible, if observations can only be made on one fault. σ 2 is perpendicular to the
direction of displacement,   found from the striations direction
for example (fig.5.19).

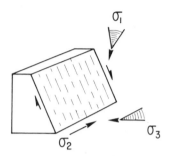

Fig.5.19. Approximate directions of the normal principal stresses
on a fault of known movement.

Where an anisotropic rock has a cleavage or is already
affected by a network of fractures, the orientation of faults may
be guided by these structures, thus only imperfectly reflecting
the state of the stresses that caused them. It then becomes
necessary to undertake analysis using statistical methods such as
have been developed in recent years. Such studies applied to
faults that have been active recently are known as **neotectonics**.

5.5.6. Complex systems-fault propagation

The general relationships between fractures, stylolitic
joints, faults and the principal stress directions are summed up
in figure 5.20. Nevertheless there are other complex systems

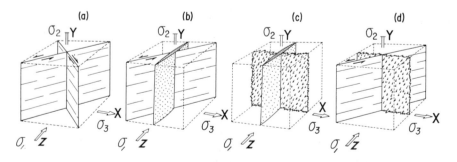

Fig.5.20. General relationship between the principal structures
of discontinuous deformation and the principal directions of
deformation (X,Y,Z) and stress ($\sigma_1, \sigma_2, \sigma_3$). a) Conjugate transcurrent faults. b) Sinistral transcurrent fault and tension frac-
ture. c) Tension fracture and stylolitic joint. d) Sinistral
transcurrent fault and stylolitic joint (after Arthaud and
Choukroune, 1972. Rev. Inst. Fr. Petr., 715).

associating different types of fractures and faults. We have
already pointed out the co-genetic association between tension
joints and faults (fig.5.8). Figures 5.15 and 5.21 also show that
the propagation of shearing in an isotropic material can be a
result of the concerted action of different types of fractures.

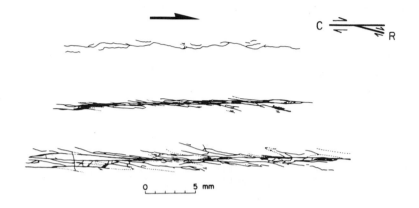

Fig. 5.21. Experimental shear in a limestone under a confining
pressure of 500 MPa, illustrating the formation of second order
shears R during increasing deformation (from top to bottom :
displacement of 0.14 cm, 0.26 cm and 0.54 cm) (Bartlett et al.,
1981. Tectonophysics, 79, 255).

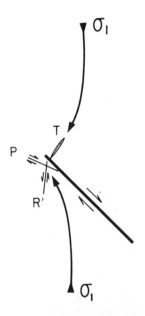

Fig.5.22. Sketch showing the rotation of σ1 trajectories in the
vicinity of a fault with the initiation of second order fractures
T, P and R' respectively on the sides under tension and compression.

The best way of understanding such associations is by a close study of the stress-fracturing relationship. Anisotropy of a medium resulting from the presence of a fault has the effect of locally modifying the distribution of stresses ($\sigma 1$ and $\sigma 3$ in particular) at its extremities, that is, in the propagating zones of the fault. In the half-plane that is under tension, the trajectory of $\sigma 1$ curves progressively and tends to become perpendicular to the fault-plane, whereas in the other half-plane which is under compression, the path of the fracture tends to curve towards the fault-plane (fig.5.22). Theoretically, the pathway of $\sigma 2$ is not affected in as much as this deformation is plane. This also explains why the lenses formed by en echelon tension fractures as shown in figure 5.5 are not parallel to the general direction of $\sigma 1$ (bisector of the acute angle formed by the two shear zones) ; their orientation being controlled by local rotation of $\sigma 1$ in the neighbourhood of the fault zones. It is probably for the same reason that shear zones end in splayed branches (fig.8.16). Thus, the formation of the main fault is accompanied by the appearance of **second order fractures** whose orientation is shown schematically in figure 5.23. These can be shear fractures or shear zones (fig.8.16b) as well as tension gashes (fig.5.8a).

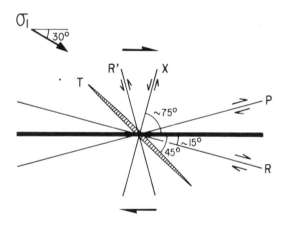

Fig.5.23. Pattern of second order fractures associated with a dextral fault (thick line).

This analysis conforms with the experimental results obtained in the shearing of an argillaceous layer. The corresponding model, according to Riedel, is shown in figure 5.23. The second order faults R and R' lie respectively at angles, measured in the trigonometric sense from this dextral main fault direction, between -(5°-25°) and +(95°-115°) to the main fault direction, -15° and +105° being average values. The tension fractures T are at approximately +135° to the fault plane. Finally, second order faults P and X may exist which are respectively symmetrical to R and R'.

These different fractures are not activated at the same time. If the external conditions cause dilation of the fault

zone,   the   P,   X,   T system is activated   preferentially
(fig.5.15a);   in compression of the zone,   it is the R, R' system
which appears . The first situation predominates if the stress σn
($\S$ 2.3.2) on the fault-plane is relatively small, where either the
confining pressure is small (as in superficial faulting),   or the
effective pressure is small (due to high fluid pressure),   or the
fault   follows a pre-existing plane of anisotropy which runs near
to $\sigma 1$ (as in figure 2.15, where $\sigma n1$ is small with respect to $\sigma 1$).
The second situation predominates on the other hand,   in deep or
dry faults , or those where the angle between the fault and $\sigma 1$ is
large.

FOR FURTHER READING

Mercier, J.L.,   1976. La néotectonique, ses méthodes et ses buts. Un
exemple :   l'arc égéen (Méditerranée orientale). Rev. Geogr. phys. géol.
dyn.,XVIII, 323.

Turcotte, D.L.   et Schubert, G., 1982. Geodynamics. John Wiley , New
York, 450 p.

Chapter 6

Structures Caused by Homogeneous Deformation

6.1. INTRODUCTION

We have seen in chapter 1 that we should be particularly interested in homogeneous deformation. In this chapter we shall study its structural expressions, and in the next one their interpretation and the processes that cause them. The concept of homogeneous deformation does not exclude the possibility of heterogeneous deformation as discussed further in § 8.1. Eventually, the imprint of homogeneous deformation is not radically different whether the medium is solid or corresponds to a crystalline suspension in a fluid ; this last case is implicitly considered here.

In a crystalline material undergoing homogeneous plastic deformation exceeding 30 % in pure strain shortening or $\gamma = 0.7$, corresponding to an angle of about 35° in simple shear , a planar and linear structural anisotropy generally develops (fig.6.1). This fabric is reinforced by increasing strain. Experiments have shown that where shortening amounts to 50 % in pure strain, or $\gamma = 1.4$ in simple shear, the trace of previous anisotropy tends to be lost. Following the state of strain (§ 2.1.3), the newly created anisotropy can correspond to a flattening type (a planar anisotropy), implying that in the anisotropy plane there is no preferred direction (K = 0 in fig.2.5), it can be uniquely linear (K = ∞), or planar-linear, the most common structure. These planar-linear structures are fundamentally the result of progressive flattening-elongation and concordant orientation of the main rock-forming minerals. They are often emphasized by a **tectonic layering** the origin of which will be discussed below.

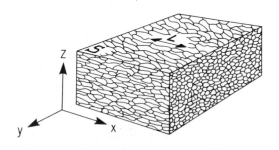

Fig.6.1. Planar-linear structure. The anisotropic surface S has a lineation L. X,Y,Z refer to the axes of finite deformation which are generally associated with such a structure.

6.2. PLANAR STRUCTURES

We shall first consider penetrative structures and then non-penetrative structures, the latter generally being emphasized by a layering or microlayering. The rocks that have a planar structure are generally fissile, that is to say that they readily split into parallel thin sheets along this plane. Their fissibility is a result of their anisotropy and corresponds to a fine and regular cleavage in slates and to a wider and more irregularly spaced parting in the case of more massive rocks such as limestones or gneisses. Sometimes these structural discontinuities act later as slip planes (fig.6.11).

6.2.1. Slaty cleavage and foliation

The principal minerals that form the rock are arranged in a platy fashion in the same plane. In the case of **slaty cleavage** the crystals of quartz, calcite and above all of sheet-silicates,

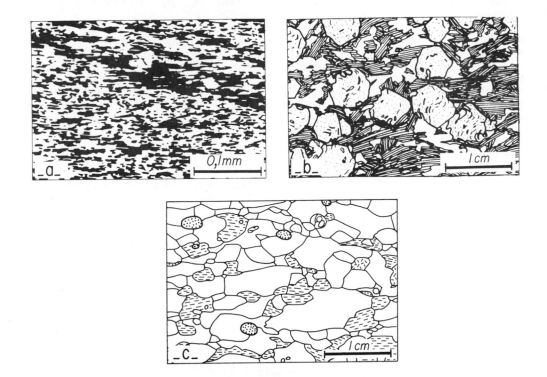

Fig.6.2. Cleavage and foliation, seen in the plane perpendicular to these surfaces and containing the lineation. a) Slaty cleavage where the contrast is emphasized by the quartzo-feldspathic minerals (white) and phyllosilicates (black). b) Foliation in a garnetiferous micaschist. c) Foliation in a peridotite. (a) and b) after Pecher, 1978. Thesis Grenoble ; c) after Boullier, 1975. Thesis Nantes).

which are mainly responsible for anisotropy, do not exceed a few
tens of microns in size whereas in the case of **foliation** the
crystals are generally visible and may attain several millimetres
in size (fig.6.2). Further to the question of size of minerals,
slaty cleavage is only found in rocks which are rich in micaceous
minerals which accounts for their high fissibility, whereas foliation can appear in a great variety of rocks (marbles,
quartzites, phyllites, gneisses, amphibolites, peridotites etc.).
In Chapter 7, we shall see that their modes of origin may be
different. Under conditions of increasing metamorphism, there is
a transition between slaty cleavage in slates and foliation in
phyllites and gneisses. **Schistosity** is a term often used to
describe the planar structure of mica-rich phyllites and gneisses.

6.2.2. Crenulation cleavage

     At the scale of a hand specimen, crenulation cleavage is
non-penetrative, and characterized by a rhythmic layering of
parallel domains spaced from a few millimetres to several centi-
metres apart (fig.6.3 and 6.4). The rock is thus **finely layered**
with layers which are often thicker, paler in colour, and richer
in quartz and/or calcite, alternating with layers which are
narrower, darker in colour and richer in micas and opaque minerals.
     Microlayers rich in quartz or calcite often correspond to
microfold hinges emphasized by a few micas. These microfolds may
be symmetrical (fig.6.3c) or asymmetrical (fig.6.3d, 6.3e).
Between these hinges, the darker mica-rich layers coincide with
the limbs of the microfolds (fig.6.3c,d). There are cases where

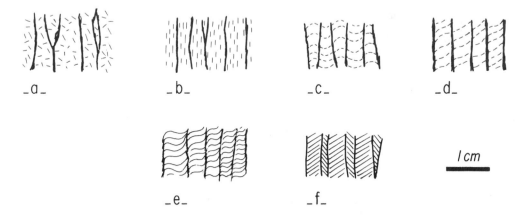

Fig.6.3. Different types of crenulation cleavage. a),b),c),d) are
cleavages marked by solution planes (shaded). a) Fracture cleavage, absence of a penetrative cleavage. b) Penetrative cleavage
parallel to the plane of concentration of insoluble minerals. c)
and d) Cleavages oblique to an older plane of anisotropy with the
development of symmetrical microfolds (c) and asymmetric microfolds (d). e) and f) Crenulation cleavages due to rhythmic microfolds with or without solution.

the rock has not undergone mineralogical differentiation and where the crenulation cleavage is simply expressed by a succession of folds with a strong common orientation along the limbs (fig.6.3e,.f). Figure 6.4 explains the origin of folds and apparent offsets seen when a planar structure of earlier origin (microlayering, cleavage, veins) undergoes solution along a plane making a high angle to this earlier structure.

Crenulation cleavage develops at a relatively low grade of metamorphism in rocks which are rich in quartz or calcite, whilst slaty cleavage develops in rocks which are richer in phyllosilicates. Transition between these two types of cleavage is frequently observed.

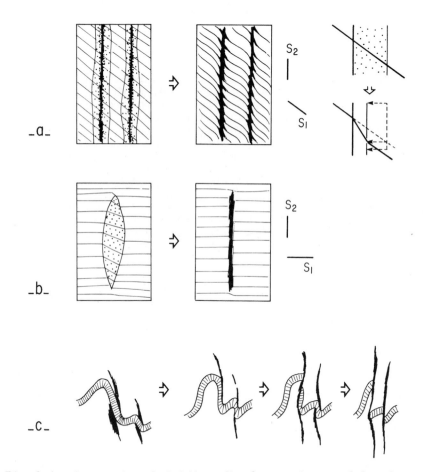

Fig.6.4. Apparent and folding displacement caused by planar dissolution (on the left initial condition, to the right present state). a) Solution along the dotted bands increasing toward the axis of the band, causing a passive rotation of S1 by shortening which gives an appearance of microfolding. b) Small kink affecting S1 ; such a fold is a preferential site for solution which is responsible for an apparent shear in S1. c) Quartz vein undergoing progressive solution, giving the appearance of displacements. (b) and c) after Gray, 1979. Amer. Jour. Sc., 279, 97).

6.2.3. Fracture cleavage

Fracture cleavage is a relatively widely spaced and irregular parting that often shows signs of solution. It preferentially
affects **competent** (that is more rigid) beds alternating with
incompetent (more ductile) beds, which develop a slaty cleavage
or crenulation cleavage. Fracture cleavage can be seen in the
hinge of open folds forming a fan converging towards the centre
of the fold (fig.6.5, § 6.2.6). The blocks created by fracture
cleavage in a competent layer are called **microlithons**. By contrast to the typical crenulation cleavage, the more rigid microlithons are not affected by microfolding. Slip can occur along
the discontinuities between **microlithons**. True slip can be distinguished from apparent slip between the microlithons due to
solution (fig.6.4) ; in the former case, the plane of discontinuity is marked by a straight fracture, possibly infilled with
minerals (quartz, calcite), and in the latter case by a progressive variation in chemical composition, for example by an
enrichment in opaque minerals or phyllosilicates.

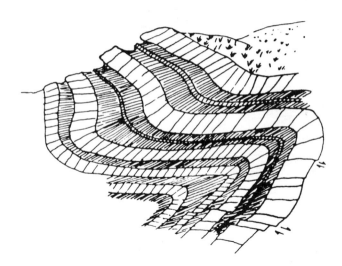

Fig.6.5. Slaty cleavage and fracture cleavage in a sedimentary
sequence of strongly contrasting competence. The slaty cleavage
is parallel to the axial plane whilst the fracture cleavage forms
fans converging towards the hinge of the folds. Note also the
decoupling of the competent beds as microlithons with apparent
slippage along the planes of weakness. (Gratier, Lejeune and
Vergne, 1973. Thesis Grenoble).

6.2.4. Tectonic layering

A layering is present at various scales in foliated metamorphic rocks which is expressed by parallel lenses and layers. The
similarity between this layering and the bedding of sedimentary
rocks can give the illusion of a direct heritage. The origin of
tectonic layering is nevertheless more complex and we shall
distinguish two types in which the origin is essentially chemical
or essentially mechanical.

Fig.6.6. Development of a metamorphic layering on a pre-existing
crenulation cleavage by the preferential growth of biotite crystals in the plane of the schistosity rich in ferro-magnesian
minerals.

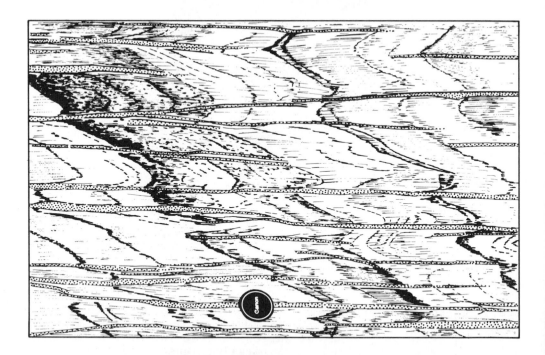

Fig.6.7. Development of a metamorphic layering due to the injection of quartzo-feldspathic magma in the axial plane of microfolds in the course of anatexis affecting granulites (photo A.M.
Boullier).

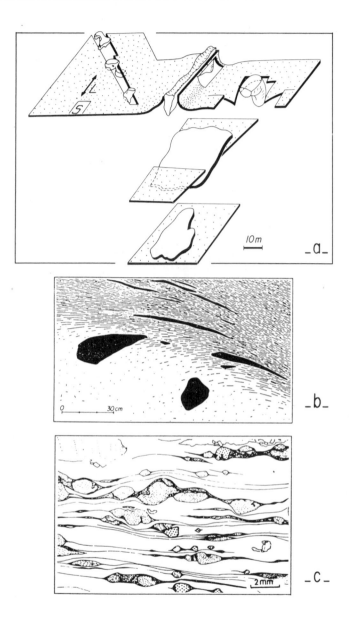

Fig.6.8. Development of layering by tectonic transposition in
zones of intense deformation. a) Progressive transposition of
irregular chromite pods in layers parallel to the foliation
(dotted plane) and elongated along the lineation L in a peridotite which was deformed at high temperature.   b) Transposition of
inclusions forming a lenticular layering in a shear zone in a
granite.   c) Transposition of pyroxene crystals partially recrystallized during a superplastic deformation forming a microlayering.   (a) Cassard et al.,   1981. Econ. Geol. 76, 805 ; b) Ramsay
and Graham, 1970. Can. J. Earth Sci., 7, 786 ; c) Nicolas and
Boullier, 1975. Phys. Chem. Earth, 9, 467).

In the first case, the rhythmic crenulation cleavage created
by a **chemical differentiation** essentially by solution within an
otherwise homogeneous rock, is preserved during increasing metamorphism and gives rise to the microlayering that is typical of
gneisses (fig.6.6). Another type of chemical differentiation is
when rocks are injected by hydrous or magmatic fluids of local or
foreign origin along parallel fractures (fig.6.7).
    All the other layerings are formed by **tectonic transposi-
tion.** This applies to rocks that were originally heterogeneous
(beds, veins, inclusions) where the heteregeneous objects by
their abundance and nature do not significantly alter the rheology of the whole rock. These objects are rotated and eventually
undergo stretching which leads them to progressively coincide
with the foliation plane. Figure 6.8 shows how inclusions of
different size and shape give rise to a lenticular layering in
zones of very large strain. In this figure the inclusions responsible for the tectonic layering or microlayering vary from chromite pods of a few metres in diameter in peridotite bodies
(fig.6.8a), to crystals of plagioclase, pyroxene or of amphibole
of centimetre size in small shear zones where their scattering is
due to superplastic deformation (fig.6.8c). In the case of beds
or veins, the rotation is rapid towards the plane of foliation in
the course of the deformation as shown in figure 6.9.

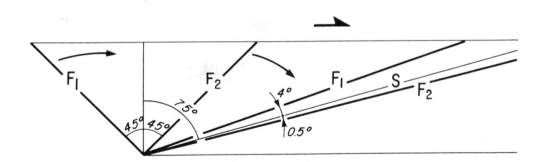

Fig.6.9. Transposition of F1 and F2 planar structures during a
shear of 75°. From their starting points, F1 and F2 end up
respectively at 4° and 0.5° on either side of the schistosity S
formed during shearing.

    An isoclinal fold which repeats the layers helps to form a
tectonic layering (fig.6.10c). Incidentally, one can commonly
observe a stretching and boudinage of layers along the limbs of
the fold into **intrafolial lenses** and a thickening in the fold
hinges due both to flow and doubling up of the layers (fig.6.10a
and b) ; when isolated in the matrix, they are called **intrafolial
hinges.** A layer or vein within the original rock can thus be
transposed into numerous layers and lenses scattered through the
fabric of the rocks.

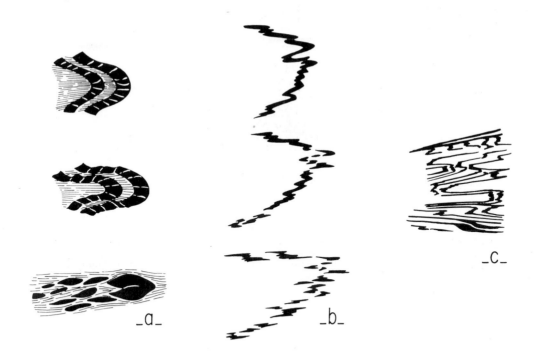

Fig.6.10. Tectonic transposition of layers during folding into a lenticular layering. a) Break-up of layers into microlithons, progressively stretched and scattered throughout their matrix. b) Splitting up of intrafolial lenses which were derived from the thickening of fold hinges and thinning of limbs during tight isoclinal folding. c) Layering being transposed by duplication of layers during isoclinal folding.

6.2.5. Cleavage, foliation and strain ellipsoid

Where rocks have a **penetrative** and **homogeneous** cleavage or foliation, markers of finite strain (fossils, reduction spots, oolites, etc.) indicate that the foliation plane contains the X and Y axes of the strain ellipsoid and that the Z axis is at a right angle to it (fig.6.1).

This conclusion is not strictly valid in rocks where the fabric is not penetrative. Thus, if there are bands of slip or solution, the deformation becomes discontinuous and the axes of the representative ellipsoid do not necessarily coincide with the structural axes (fig.6.11). The acquisition of a planar anisotropy parallel to the cleavage-foliation plane tends to facilitate slip along these planes. To be valid the strain analysis should be able to account for such slip discontinuities . The coincidence of the X,Y,Z axes with the structural axes is not certain in the case of fracture or crenulation cleavage, because of their discontinuous deformational character. The question of the coincidence of the two sets of reference axes can only be re-examined on a spatial scale where the deformation can be considered to be penetrative and homogeneous.

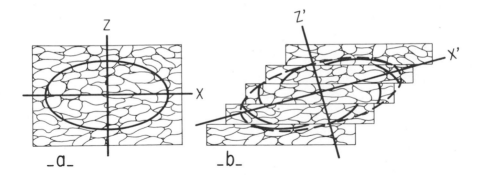

Fig.6.11. Relationship between foliation and the strain ellipsoid
in the plane perpendicular to Y. a) Usual situation where Z
coincides with the normal to the structural plane and X with the
lineation. b) Situation where a discontinuous slip takes place
along the plane of foliation : the strain ellipsoid is changed in
shape and orientation (dashed outline and new axes X',Z').

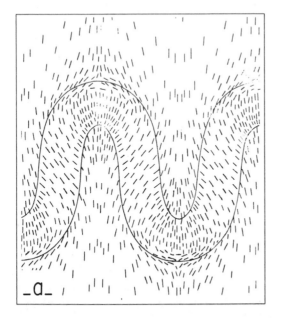

Fig.6.12. Cleavage and trajectories the X axis of the strain
ellipsoid in layers of different viscosities. a) Numerical simulation for a viscosity ratio of 42/1 between the folded layer and
its matrix, stress applied E-W. b) For comparison, microfold in
quartzites surrounded by slates (Dieterich, 1969. Am. Jour. Sci.,
267, 155).

6.2.6. Cleavage, foliation and folds

When a rock is folded, the cleavage and foliation are generally parallel to the **axial plane** of the folds (§9.5.1, fig.6.5,
6.10a), especially if the folded layers can be equated with
passive markers, as when their viscosity is close to that of
their surroundings. By contrast, when the folded body has
substantial differences in viscosity, such as in interstratified
limestone-marl or sandstone-shale, the nature of the cleavage as
well as its orientation varies from layer to layer. Thus in
competent beds (limestone or sandstone) a fracture cleavage appears which fans out from the centre of the fold, and by contrast
in the incompetent interbedded horizons (marl or shale), a slaty
cleavage is developed which is parallel to the axial plane or
forms a slightly diverging fan, (Fig.6.5, 6.10a and 6.12). This
is termed **cleavage refraction**. The cleavage trajectories and in
particular its refraction on passing from a competent bed to an
incompetent one are reproduced in the experimental folding of a
more viscous bed in a medium of lower viscosity ; the trace of
the large axis X of the finite strain ellipsoid follows that of
the schistosity quite faithfully (fig.6.12). This close coincidence shows that in a situation as complex as that of the folding
of a stratified medium of variable viscosity, the cleavage hardly
deviates from the plane containing the X direction of the strain
ellipsoid.

6.3. LINEATIONS

The term **lineation** is given to all linear traces upon the
plane of cleavage or foliation (fig.6.13). This reference to the
foliation or cleavage is indispensable as a line upon another
plane, with the exception of layering in some particular cases,
has no significance. In practice, one first of all identifies the
cleavage or foliation plane before looking for the lineation on
that plane. This subordination and the intrinsic difficulty of
observing and identifying lineations in the field perhaps explains why they have been neglected in some studies. The structural frame is then incomplete, which introduces an obvious limitation to structural studies. There exist nevertheless formations,
called "L-tectonites (as opposed to "S-tectonites" where only the
foliation is identifiable), in which a plane cannot be observed
but only a lineation because of the axial symmetry of the deformation in relation to this lineation (K=∞ in § 2.1.3). It is
also essential to identify carefully the nature of the lineations
mapped in the field . A planar structure can have several lineations stemming either from a single homogeneous deformation which
often causes orthogonal lineations (fig.6.13a), or from superimposed deformations (fig.6.14). We shall describe the principal
types of lineations illustrated in fig.6.13.

Finally we note that fractures either dry or filled with
fibrous minerals are often perpendicular to the lineations. If it
is confirmed that they are tension fractures formed at the same
time as the lineations (by looking at the nature of their infilling), such fractures can help to identify perpendicular lineations as stretching lineations (parallel to the X axis of strain
ellipsoid).

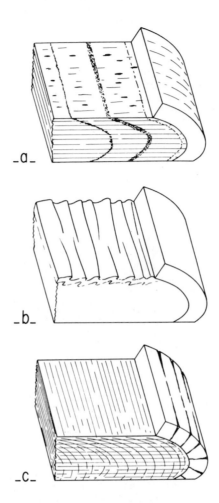

Fig.6.13. Principal types of lineations. a) Intersection lineation, between a folded layer and an axial plane cleavage, seen in
the plane of cleavage as linear traces of different compositions
parallel to the fold axis. In the same plane there is also at
right angle, a mineral or aggregate lineation ; upon the folded
layer, striae parallel to the latter mineral lineation. b) Corrugation lineations parallel to microfold axes. c) Intersection
lineations : intersection of two cleavages observed on the axial
plane cleavage, and intersection of the folded layer with a
fracture cleavage observed on the surface of the folded layer.

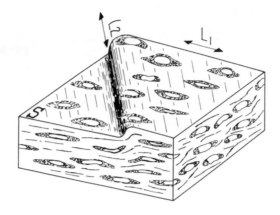

Fig.6.14. Superimposition of two independent lineations : L1, mineral and aggregate lineation ; L2, corrugation lineation.

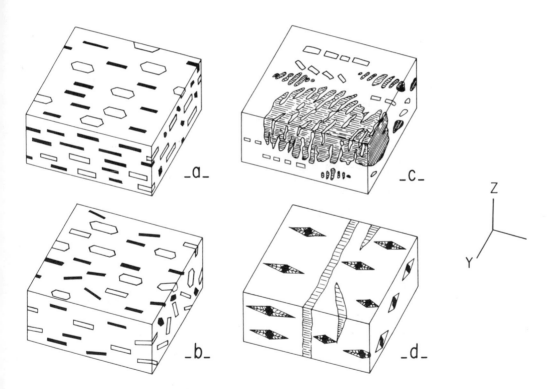

Fig.6.15. Mineral and aggregate lineations. a) Mineral lineation formed by alignment of acicular and tabular minerals. b) Mineral lineation due to alignment of acicular minerals and zonal dispo- sition of tabular minerals. c) Pull-apart lineation affecting minerals or aggregates. d) Lineation formed by infillings of fibrous minerals in veins or around hard objects in pressure shadows.

6.3.1. **Mineral lineations**

Mineral lineations are characterised by a parallel orientation of minerals having an anisotropic fabric. Minerals may have
a prismatic or acicular habit as in the case of amphibole, pyroxene, sillimanite, tourmaline or a tabular habit as in phyllosilicates, chloritoid, some feldspars, etc. In the latter case, the
lineation is either due to a preferential elongation of the
tabular faces (fig.6.15a), or to the arrangement of these minerals around an axis which defines the lineation (fig.6.15b). In
the course of a large deformation, some aggregates or minerals
which are less ductile than their matrix fracture at right angle
to the direction of stretching (fig.6.15c). The matrix squeezed
into the fractures divides them into isolated fragments. The
normal to these tabular-shaped fragments coincides with the direction of stretching. This **pull-apart** lineation is well known in
orthogneisses and peridotites where it affects respectively alka-
li feldspars augen and orthopyroxene crystals or massive chromite
aggregates. Lastly, one can include in the category of mineral
lineations **fibrous infillings** of fractures or of pressure shadows
(fig.6.15d, § 8.2.1).

6.3.2. **Aggregate lineations**

An aggregate lineation is so called when objects distinct
from the matrix are aligned in the cleavage or foliation plane
(fig.6.14). When metamorphism is weak or absent, the origin of
the objects can be determined ; for example lineations due to
stretched fossils, pebbles, phenocrysts, etc. When metamorphic
grade is high, recrystallization produces aggregates of crystals
which are defined by their composition, for example "quartzo-
feldspathic aggregates". **Rods** are lineations formed of aggregates
which are often quartz-rich and very elongated, forming parallel
bundles (fig.6.16a). The rodding structure is common in facies
where the foliation is difficult to identify.
During bed over bed slip, hard objects can inscribe their
imprint upon these beds in the form of **striations** which are
parallel to the slip direction (fig.6.13a and 5.17d).

6.3.3. **Intersection lineations**

Intersection lineations appear each time that a new cleavage
or foliation is superimposed obliquely upon an older bedding or
cleavage, as when the new cleavage lies in the axial plane of the
fold deforming the bedding (fig.6.13a) or an older cleavage
(fig.6.13c). Intersection lineations can be seen in the axial
plane cleavage as well as in the deformed bed. In the cleavage
plane it appears as distinct parallel lines resulting from the
intersection of deformed beds and the cleavage plane. When the
bedding is fine-grained and more or less lenticular, this type of
lineation is easily confused with rodding (fig.6.16b) ; a close
examination is then necessary as the significance of the two
lineations is generally very different. Intersection lineation
can also be seen in the deformed bed. It then appears as the
trace of the axial plane cleavage. When the folded bed was alrea-

dy schistose as in the case of ancient marls or pelites, the intersection of this former cleavage with the new axial plane leads to the break-up of the rock into small blocks or prisms, defining a **pencil lineation**.

6.3.4. Corrugation lineations

This type of lineation is due to microfolds with their axes more or less parallel to each other (fig.6.13b). The general appearance is that of corrugated iron. The microfolding that is responsible for the corrugation lineations is produced in fine grained incompetent materials such as pelites, marls, phyllites and some micaschists. The competent beds develop **mullions** sometimes at the contact with incompetent beds which differ from corrugations by their larger wavelength, greater longitudinal regularity and a scalloped profile, such that the convex side faces the incompetent bed. The asymmetric profile is due to the contrast in viscosity between the competent and incompetent beds.

6.3.5. Lineations and the strain ellipsoid

The relationship of the principal X and Y axes of the strain ellipsoid to the orientation of lineations cannot be answered immediately. Lineations related to folds can provisionally be removed from the discussion (see § 6.3.6 and 9.5.2). We quote here a few of the lineations which tend to lie parallel to the X axis of the strain ellipsoid :
    i) mineral lineations formed by prismatic minerals such as amphiboles, or by more platy minerals which are nevertheless elongated in one direction such as feldpars ;
    ii) lineations due to recrystallization in tension fractures, sheltered zones and pressure shadows ;
    iii) aggregate lineations, where they can be shown to result from the re-orientation of previously elongated objects or the stretching out of objects of more equiaxial shape.
    Contrary to the preceding case, lineations which have been formed by the break-up of layers by stretching, tend to be coincident with the Y direction of the strain ellipsoid. This is the case where stretching of competent layers in an incompetent matrix produces **boudinage** of these layers (fig.6.16c and § 8.4) or a mineral lineation formed by "pull-apart" (fig.6.15c). The following examples underline the importance of a good diagnosis concerning the nature of lineations before coming to conclusions about their relationship to the principal axes of finite strain.
    When strain is intense and non-plane, a boudinaged layer undergoes a minor degree of stretching in the Y direction. The boudins then adopt a spindle-shaped form parallel to Y (fig.6.16c). This boudinage acting upon a layer of an appropriate thickness, can make a rodding structure that is easily confused with that formed by stretching a conglomerate. In this latter case the rod lineation is parallel to X. A rod lineation can also be formed by the break-up of a folded layer, often quartz rich, by a cleavage or a foliation ; this rodding may be parallel to the X or Y directions. In favorable instances analysis of intracrystalline plasticity throw light upon these doubtful cases.

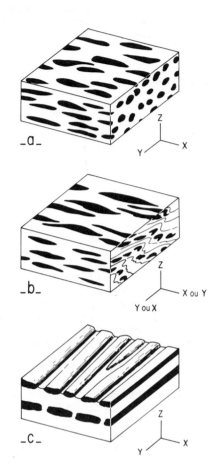

Fig.6.16. Relationships between rodding and the principal axes
X,Y,Z. a) Rods formed by stretching isolated objects such as
pebbles in a conglomerate. b) Rods formed by the disruption of
intrafolial lenses formed by folding. c) Rods formed by boudinage
of a competent bed.

6.3.6. Lineations and folds

It is unfortunate that for too long lineations have been
analysed through their relationship to folds (§9.5.2). As a
result of some unfounded interpretations regarding the kinematics
of these folds, incorrect interpretations of the associated lineations have been proposed.

By their nature, intersection and corrugation lineations are
parallel to the fold axes. The same applies to lineations formed
by a zonal arrangement of tabular minerals (fig.6.15b), when they
are associated with folding. Mineral or aggregate lineations
produced by stretching (parallel to X) often make a significant
angle with the fold axis, in areas of moderate strain.

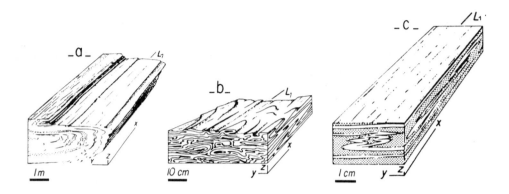

Fig.6.17. Lineations parallel to fold axes. a) Intersection and corrugation lineations. b) Corrugation lineations parallel to sheath folds (§9.2.1). c) Intersection and mineral lineations. (Mattauer et al., 1981. Jour. Str. Geol., 3, 401).

Contrarily, they are often parallel to the fold axis (fig.6.17) when strain is large (fig.6.17).

We have seen how the axial plane of folds can coincide approximately with the YX plane of the finite strain ellipsoid. The question posed by the angular relationships between linea- tions and the fold axis then becomes that of knowing in which direction the fold axis lie in the plane of YX. We shall see in that Chapter 9 that the answer is not unique ; as a result the question of interpreting intersection, corrugation lineations and those defined by zonal arrangement of minerals depends on the interpretation of the associated folds.

FOR FURTHER READING

Borradaile, G.J., Bayly, M.B. et Powell, C.McA., 1982. Atlas of deforma- tional and metamorphic rock fabrics. Springer-Verlag, Berlin, 551 p.

Cloos, E., 1962. Lineation, a critical review and annoted bibliography. Geol. Soc. Amer. Mem. 18, 122 p.

Turner, F.J. et Weiss, L.E., 1963. Structural analysis of metamorphic rocks. McGraw Hill, New York, 545 p.

Chapter 7

Interpretation of Continuous Homogeneous Deformation Structures

7.1. INTRODUCTION

In chapters 4 and 6 we have considered in turn the mechanisms and then the results of continuous homogeneous deformation. By identifying the deforming mechanism responsible for the structure studied and by means of complementary data, one can attempt to define the nature of the movements (kinematics) and/or stresses (dynamics) that gave rise to the structure.

In an account of structures formed by continuous deformation, one often ignores those which have been formed by the flow and all the more so by static deposition in a **fluid**. The fluid may be a suspension of solid particles in water or of crystals and inclusions in a **magma**. These structures can also be interpreted in a kinematic and dynamic manner. Thus the study of fluidal structures in a granitic massif allows one to define the geometry of the massif in three dimensions, to work out the kinematics of its intrusion and to consider the forces which produced it (fig.7.1).

Some of the structures produced in a fluid medium are remarkably similar to those produced in the **solid state** (fig.7.4), a brief description of criteria characterising one from the other is indispensable. That is why in this chapter we try firstly to

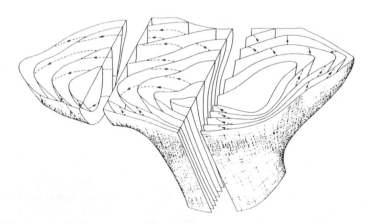

Fig.7.1. Magmatic flow structures in a granodioritic batholith, many kilometres across. The partitions correspond to the planes of flow and the dashed lines to the directions of flow. (Pons, 1982. Thesis Toulouse).

define the criteria of different modes of flow before looking   at
the   kinematic   and dynamic significance of the   structures   they
produce.

7.2. MECHANISMS RESPONSIBLE FOR THE DEVELOPMENT OF STRUCTURES

7.2.1. Deposition and flow in a fluid

Sedimentation   of   particles   and crystals,   or   their   flow
within a fluid medium (water or melt),   leads to the   development
of planar-linear structures. Thus the rhythmic character of sedimentation   creates   a bedding upon which compaction may impose   a
cleavage. When   the   sedimentation is   dynamic   (current   marks,
slumps),   the   plane   of deposition often has a lineation and   is
folded in the same way as a rock deformed in the solid state.
The   nature   of   the rock and the geological   context   often
allows one to distinguish whether the structures were formed in a
solid or liquid medium.   Thus the fluidal structure of a lava   is
attributed   unequivocally to magmatic flow (fig.7.2).   Similarly,
in   a granite,   the magmatic structure shown by the alignment   of
feldspars and ferromagnesian minerals (fig.7.3a) is usually easily   distinguished from foliation produced by plastic   deformation
in   the solid state (fig.7.3b,   see   also § 7.2.2).   Nevertheless
there   are more ambiguous situations,   thus folds in   sedimentary
rocks may be produced by a turbulent flow within a mass of uncon-
solidated   mud   and are then called **slumps** (fig.7.4a),   or   by   a
deformation of tectonic origin produced by plastic deformation in
the   solid   state   under deviatoric stresses (ch.9).   It is   true
that a diagnosis of tectonic deformation by solution-crystallization   is sometimes difficult to make as it is a   mechanism   which
occurs   in sediments rich in   interstitial fluids.   In the body of

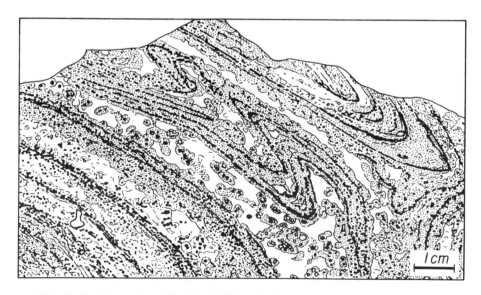

Fig.7.2. Magmatic fluidal flow structures in a rhyolite.

more or less consolidated sediments,   there can also be a   continuity   between conditions of dynamic sedimentation,   and those of
deformation in the solid state.   In this case the distinction   is
meaningless.

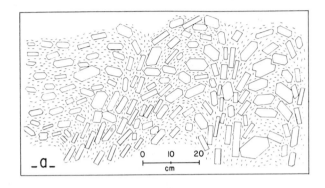

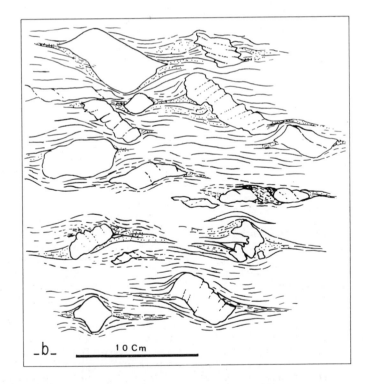

Fig.7.3. a) Magmatic   structure in a granite ;   here   the   flow
planes are folded.   The feldspars are undeformed and euhedral. b)
Plastic deformation structure in an orthogneiss derived from
granite.   The feldspars form fractured and stretched   augen.   (a)
Fisher, 1957. Aufschluss, 6,7 ; b) Brunel, 1983. Thesis Paris).

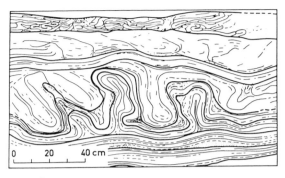

_a_

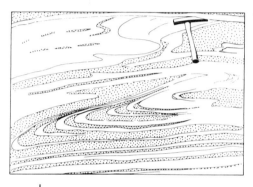

_b_

Fig.7.4. Slump structures. a) In evaporites. b) In sedimentary layers in a gabbro magma chamber. (a) Kuene, 1968. Tectonophysics, 6, 143 ; b) From a photograph by J.F. Violette).

On an outcrop scale one looks for certain typical sedimentary structures such as discordances between beds (cross-bedding, wash-outs), current structures (ripple marks, flute casts, etc.) and folds (slumps) (fig.7.4a). These structures are also known in some basic eruptive rocks, the magmatic cumulates produced by a dynamic deposition of crystals in a magma chamber at around $1200°C$ (fig.7.4b).

On a thin section scale, sedimentation in a stationary fluid or in a flow induces preferred orientation of minerals of aniso-metric habit such as phyllosilicates or some feldspars (§ 7.3). Contrary to the case of plastic deformation, in a fluid the preferred orientation of minerals results solely from their shape, as illustrated by the planar deposition of plagioclase tablets in a magma (fig.7.5). In effect the considerable contrast in viscosity between the fluid and the crystals means that there is no intracrystalline deformation. In the same way, during slump folding, minerals tend to wrap around the hinge without under-going deformation (fig.7.3a).

layers have different velocities (we say that there is a **velocity gradient**) (fig.7.10), which creates shearing between the layers. The structural plane produced by laminar flow is called the plane of **lamination**. The case of **turbulent** flow causing slumps (fig.7.4 a and d) and apparently disorganised structures (fig.7.3a) will be considered with folding in chapter 9. We shall examine firstly the reorientation of deformable objects, then that of particles and rigid objects.

An eruptive rock often contains deformed inclusions (fig.7.7) which are due to the imperfect mixing of magmas. If that is the case, the problem of deformation and of the reorientation of these inclusions can be considered as that of the deformation of passive markers, that is to say markers having a comparable viscosity to that of the surrounding medium and which do not alter its mechanical behaviour. Such markers progressively lengthen in the (X,Z) plane of deformation parallel to X and permit a precise measurement of the finite deformation when their viscosity approaches that of the surrounding medium (see Appendix II). However this is not the case when viscosity of the marker is much greater than the medium ; the situation becomes one of rigid particles, a case now described.

Fig.7.7. Deformed inclusions within a diorite. Such inclusions are produced by the imperfect mixing of two magmas and they are deformed as passive markers (Pons, 1982. Thesis Toulouse).

Rigid particles acquire a preferred orientation when they have an anisotropic shape (anisometry). In view of the torque that exists between two adjacent layers of the fluid, the anisometry plane of these rigid objects undergoes a progressive rotation which leads to their alignment in the flow plane (fig.7.8). Once acquired, this orientation is stable if the flow is coaxial. On the contrary in the case of a non-coaxial flow, this orientation is not generally stable. If the orientation of the anisometry plane of the object goes through the flow plane, the torque that then results will cause a new turn-over (fig.7.8). The rate of rotation depends upon the degree of anisometry of the particules and upon the instantaneous angle between the long axis of the object and the flow plane. For particles which do not

interfere and provided the flow has been large enough ($\gamma \geqslant 1.5$),
the preferred orientation of the anisometry planes corresponds in
a permanent regime to a small angular domain around the flow
plane. If particles are numerous, this preferred orientation is
more difficult to attain by the more anisometric particles which
can be in contact and tend to linger in the plane of finite
deformation. When they are particularly abundant the anisometric
particles in suspension interact strongly in the course of their
rotation and imbricate like tiles on a roof, this phenomena is
tiling (fig.7.9). Finally if the particles are also elongated,
their long axis tends in the same way to align itself
statistically with the direction of flow or of finite deformation
because a velocity gradient can exist in the flow plane itself.

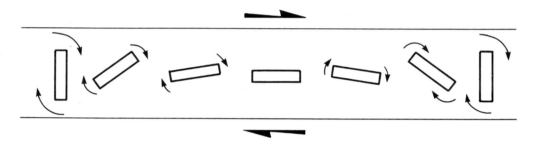

Fig.7.8. Rotation of a rigid tablet during the shearing of its
matrix. The length of the arrows is proportional to the rate of
rotation.

    Thus, thanks to a statistical analysis of the planar and
linear orientations of particles and crystals, one can know the
orientation of the plane and lineation of viscous flow. If the
geometry of flow is known independently, as in the case of flow
upon an identified substratum or in the interior of a dike of
which the walls are known, tiling allows the definition of the
sense of flow (fig.7.9).

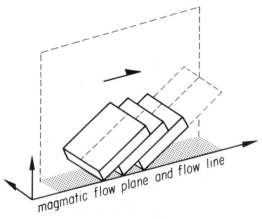

Fig.7.9. Mutual superimposition of rigid tablets (tiling) in a
shear flow (Den Tex, 1969. Tectonophysics, 7, 457).

    In fluid flowing along a wall, there is a very high velocity
gradient between layers as the velocity, which is zero along the
wall, increases rapidly away from it. If the fluid contains
particles in suspension, the mechanical interactions between them
are particularly strong near to the walls. Thus in this area of
high gradient a dispersive pressure appears which tends to move
the particles towards the zones of zero velocity gradient, that
is to say away from the wall (fig.7.10). This effect, called the
Bagnold effect, causes a segregation of particles in the fluid if
their concentration exceeds 5 % and if the velocity gradient is
sufficiently large ; this corresponds geologically to dikes and
intrusions less than several tens of metres in width. Some magmatic banding can be attributed to the Bagnold effect.

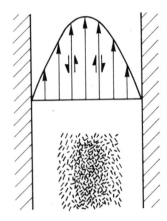

Fig.7.10. Flow between the walls of a fluid charged with
particles. The particles concentrate in the zone where the velocity gradient is weak or zero.

7.4. INTERPRETATION OF PRESSURE-SOLUTION DEFORMATION STRUCTURES

7.4.1. Relation between cleavage-lineation and principal stresses

    We saw in chapter 4 that cleavage or foliation produced by
the pressure-solution mechanism tends to appear parallel to the
principal stress directions : the cleavage perpendicular to the
maximum principal stress $\sigma 1$ and the stretching lineation parallel
to the minimum principal stress $\sigma 3$ (fig.4.21 and 6.12). Such a
rule would be very important as theoretically it would mean that
the principal stress directions responsible for a cleavage-
stretching lineation couple could be determined. It applies unre-
servedly when the progressive deformation is coaxial, as in the
diagenesis of sediments by compaction in the course of burial or
in simple stylolitic joints where the peaks point in the direction of $\sigma 1$ (fig.5.9). Tectonic deformation is on the contrary
frequently non-coaxial. The deformation axes rotating in relation
to those of the principal stress during progressive deformation
(fig.2.17b), the situation then becomes more complex. In view of

the preceding rule, for each increment of deformation the cleavage acquired should reorientate perpendicular to $\sigma 1$. If that
were the case, the cleavage would remain permanently normal to
the $\sigma 1$ stress direction and would rapidly cease to coincide with
the (XY) plane of the ellipsoid of finite deformation which
progressively rotates towards the direction of the shear plane.
We have seen (§ 6.2.5) that the cleavage plane generally coincides with the (X,Y) plane of finite deformation. Then the cleavage plane rotates and is only perpendicular to $\sigma 1$ during the
first increments of the deformation (see § 2.5). Let us assume
that these first increments have already induced a notable
planar-linear anisotropy. Pressure-solution would then operate in
relation to this new orientation under the dependence of the
stress component σn which is perpendicular to this early cleavage
and not of the applied $\sigma 1$ stress (fig.2.15). Cleavage would thus
follow the imposed rotation, and thus it would be less dependent
on the principal stresses than on the presence of planes permit-
ting an easy circulation for fluids and allowing a more rapid
diffusion of chemical elements.

7.4.2. Origin of crenulation cleavage

We have seen that crenulation cleavage could be expressed by
a rhythmic succession of microfolds without showing signs of a
particular mineralogical differentiation (§ 6.2.2). Its origin is
then entirely mechanical, resulting from compression or
microshearing (fig.8.16c) upon a rock that already possesses a
good slaty cleavage.

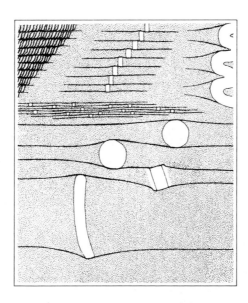

Fig.7.11. Spacing of solution planes, controlled by the
distribution of insoluble objects (white) : inclusions of various
shapes, regularly distributed crystals, broken objects, microfolds etc. (Gratier, 1984. Thesis Grenoble).

On the contrary one must attribute crenulation cleavage   to
a pressure-solution mechanism when there is a chemical differentiation expressed by a residual concentration of insoluble minerals (fig.6.4).   The rhythmic and finely penetrative character of
the solution surfaces is poorly understood.   It could be controlled by the shape and distribution of pre-existing heterogeneities
(microfolds,   inclusions or insoluble crystals,   etc.)   (fig.7.11
and   6.4b).   If   an external stress is applied there is a   stress
concentration   at   the contact of these   insoluble   objects   that
could   start   the process of solution.   Solution would then extend
on a surface perpendicular to $\sigma 1$,   or corresponding to a circulation channel for fluids ; this last interpretation could apply to
the solution surfaces shown in figure 6.4a.

7.4.3. Origin of fracture cleavage

Fracture   cleavage (§6.2.3) commonly forms in competent   and
massive   beds by the same solution mechanism which forms crenulation   cleavage in incompetent ones.   Comparison   with   stylolitic
joints then becomes necessary (§ 5.4).   The microlithons formed by
the   fracture cleavage may show signs of tangential   displacement
or of opening with infilling. These movements prove that fracture
cleavage   may   have a complex history and perhaps other   possible
origins.   The fan-like arrangement of fracture cleavage with respect   to the axial plane of folds (fig.6.5 and 6.12)   could   thus
result either from an orientated solution controlled by a locally
deviated principal $\sigma 1$ stress (fig.5.22) (fig.7.12a)   or from   the
rotation   of   a solution cleavage formed at an earlier   stage   of
folding (fig.7.12b).

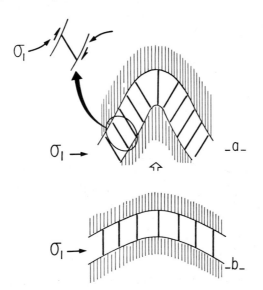

Fig.7.12.   Possible origins of a fracture cleavage fan.   a) Solution guided by local deviation in the $\sigma 1$ trajectory.   b) Solution
cleavage   formed   during   an earlier stage of folding and   later
transposed passively.

7.5. INTERPRETATION OF INTRACRYSTALLINE PLASTIC DEFORMATION STRUCTURES

Plastic deformation by intracrystalline slip produces in a given crystal a shear at the macroscopic scale as shown in figure 4.10. It is thus a non-coaxial deformation which is shown by a progressive rotation of the representative ellipsoid of deformation in the direction and amount of the slip. Preferred orientation of an aggregate which develops during progressive deformation is the summation of the individual reorientations of the constituent crystals under plastic deformation (fig.7.13).

7.5.1. Development of preferred orientations

Plastic deformation has the effect of simultaneously developing shape and lattice preferred orientations. The **shape preferred orientation** ("S" on figure 7.13) is the expression of an average orientation of flattening (foliation) and elongation (lineation) directions of the crystals. The **lattice preferred orientation** ("C" on fig.7.13) is the expression of the average crystallographic orientation of the crystals which develops during plastic deformation by a mechanism that we shall look at briefly.

We have seen that deformation by slip in a crystal is accompanied by rotation of all the lines which are associated with it (save those in the slip plane) (fig.4.10). In the coaxial deformation of an aggregate represented in fig.7.14, we look at an individual crystal. In order to adapt to the deformation of the aggregate it has to extend by shearing and its Xc axis undergoes a rotation of + d β c, due to shearing, for each increment of deformation. This crystal has to undergo a simultaneous global

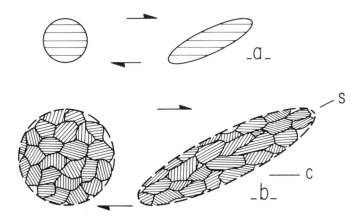

Fig.7 13. Comparison between deformation of a) a monocrystal in an orientation amenable to slip and b) an aggregate having crystals whose slip planes are initially in a random orientation. By plastic deformation, these planes are given a preferred orientation C, distinct from the average direction of stretching S of the constituent crystals.

rotation -d β g to preserve the alignment of its Xc axis with the
Xa axis of the aggregate. It is this last rotation which reorientates the lattice in a way such that the slip plane progressively
approaches the direction Xa.

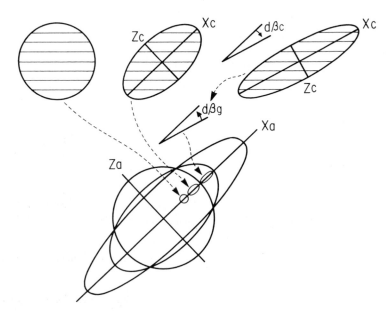

Fig.7.14.   Progressive reorientation of the slip plane (by
increment of dβg) of a crystal forming part of an aggregate
undergoing coaxial deformation.

     A two dimensional simulation permits one to follow the
development of the preferred orientation and underlines the role
played by the original orientation on the fate of the crystal
(fig.7.15). We illustrate this with the progressive coaxial
deformation of cells, which represent crystals, possessing a
single slip line. It confirms that a system with only one slip
direction, even if it is accompanied by sufficient rotations and
translations, is not sufficient to produce a homogeneous deformation in which each cell deforms in a similar manner to that of
the aggregate. Voids and overlaps appear which the simulation
tries to minimize. The amounts of rotation of the slip line and
of deformation of a cell depend upon the initial orientation of
the slip line.
     On figure 7.15a the shape of cell (a) changes but remains
pretty near to the mean shape of the aggregate and its slip
direction undergoes a large rotation. On the contrary cell (b) is
blocked ; it cannot deform or undergo rotation. Finally, the
rotation of the slip lines of the individual cells form two
symmetrical orientations which progressively move towards the
elongation direction (fig.7.15b).

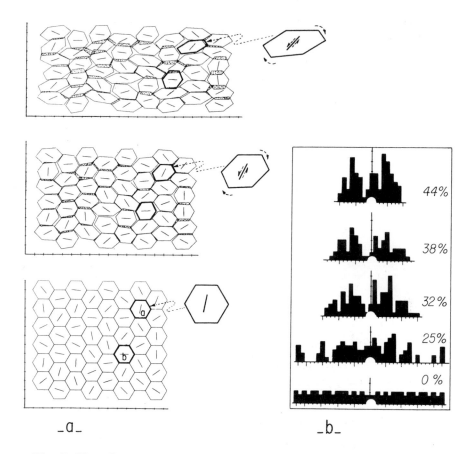

_a_   _b_

Fig.7.15.   Two-dimension simulation of progressive coaxial deformation   of   cells possessing a single slip   direction.   a)   Model
showing two cells a) and b) respectively corresponding to   orientations   favourable and unfavourable for easy slip.   b) Preferred
orientation of slip directions with increasing strain. The vertical scale graduated in %   strain corresponds to the direction   of
elongation (Etchecopar, 1977. Tectonophysics, 39, 121).

7.5.2. Heterogeneous deformation of crystals

The   two   dimensional   simulation of fig.7.15 shows that   the
working   of   a   single slip line cannot   insure   the   homogeneous
deformation of an aggregate,   in conformity with Von Misès criterion (§4.2.3). In two dimensions it is always necessary to have a
second line as shown in fig.7.16b.   This figure illustrates   that
after   the activation of a dominant slip system (central diagram)
two   situations may develop.   If the crystal only activates   this
system,   a global rotation is necessary in order to preserve   its
alignment with its neighbours (fig.7.16a), in confirmity with the
analysis in the preceding paragraph.   If a second system is activated,   the   global rotation takes place without rotation   of   the
lattice (fig.7.16b).

In three dimensions the rotation of the lattice accompanying the deformation and responsible for the preferred orientation results from the fact that natural crystals favour a restricted number of systems rather than activating the five systems which would allow accommodation of the deformation without rotation of the lattice. Thus, in figure 7.16, path (a) is followed in preference to path (b).

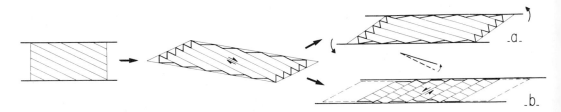

Fig.7.16. Non-coaxial deformation of a crystal between horizontal platens. The shearing on an oblique plane produces a rotation that must be compensated by : a) an equal and inverse body rotation or b) the activating of a second slip system with no lattice rotation.

At low temperature, natural crystals glide for the greater part under a restricted number of systems, such as (0001) <a> in the case of quartz deformed at low temperature. This results in a particularly heterogeneous deformation. The crystals unfavourably orientated for slip are freed by the sometimes considerable rotations of the lattice (fig.8.18h), by kinking or twinning. Syntectonic recrystallization can aid these rotations.

At high temperature, the degree of freedom of the crystal is increased thanks to diffusion which allows dislocation climb and grain boundary migration. By this last process, crystals in unfavourable orientations are consumed and eventually disappear (fig.7.17). These can also be removed by dynamic recrystallization. Such natural selection reinforces the effects of individual rotations in the development of preferred orientation.

7.5.3. Coaxial and non coaxial deformation

In a coaxial regime of deformation, the principal slip planes of crystals progressively rotate away from the $\sigma 1$ stress direction, the slip directions simultaneously approach the $\sigma 3$ direction. Figure 7.15 illustrates this evolution and underlines the symmetry of the preferred orientations with respect to the principal directions of deformation which in this case coincide with the stress directions.

In a regime of non-coaxial deformation, this symmetry is again present in the first stages of the deformation, but it disappears rapidly if deformation continues. The crystals slip with a shear sense which conforms to the imposed shear sense or in the opposite sense depending on the initial orientation of their dominant slip system. The crystals slipping in the opposite sense are rapidly blocked. This asymmetry is shown in fig.7.17 which represents the fabric evolution of ice crystals during

experimental deformation by shear. In this experiment a preferred
orientation appears at a shear angle of 35° and, beyond that,
increases progressively. The orientation of the dominant slip
system in well orientated crystals coincides statistically with
the imposed shear ; crystals in an unfavourable orientation for
easy slip are removed by grain boundary migration. This enables
the fundamental rules for kinematic analysis to be formulated.

1) During homogeneous deformation of a rock composed of
minerals possessing a dominant slip system, the preferred orientation of the slip plane and slip direction tend to coincide
respectively with the plane and direction of plastic flow.

2) The plastic flow regime can be deduced by comparing the
preferred orientation of slip systems with the directions of
finite deformation.

In coaxial deformation, these two reference frames coincide
(fig.7.15). On the contrary, during non-coaxial deformation they
are distinct ; according to the sense of rotation from one reference frame to the other, the shear sense is dextral or sinistral
(fig.7.18). Lastly it must be stated that this method of kinematic analysis is fully applicable only in the case of homogeneous
plastic deformation of large finite strain.

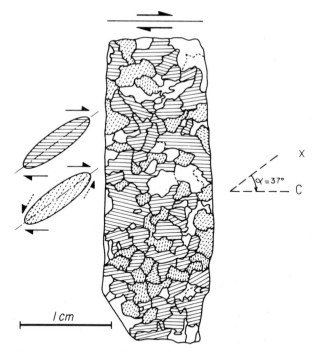

Fig.7.17. Experimental shear deformation of a test sample of
polycrystalline ice. After a shear angle of 35°, two principal
populations of crystals can be distinguished where the slip
planes (0001) are symmetrically inclined in relation to the
direction of elongation (X) : crystals in an easy slip orientation, where the slip plane (hatches) approaches the direction of
imposed shearing and crystals unfavourably oriented, where the
slip plane (dashes) is away from this direction. The former grow
relatively to the latter as deformation increases. (Bouchez and
Duval, 1982. Text. and Microstr.,5,171).

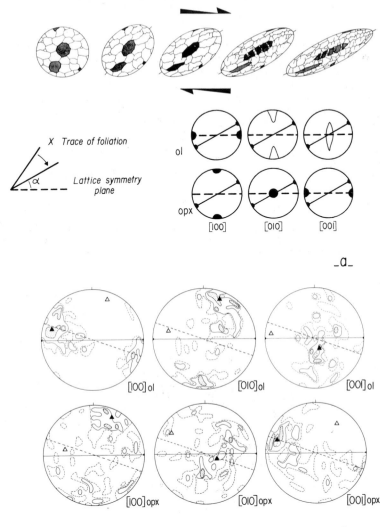

Fig.7.18.   Example of kinematic analysis in   a   peridotite.   a)
Theoretical diagrams. (olivine (ol) : white with (100) subboundary traces ; orthopyroxene (opx) : hatches parallel to the (100)
plane. With increasing strain by dextral simple shear, the trace
of foliation X rotates and is lengthened.   The stereonets corresponding to the final stage, show that the orientation of slip
systems coincides with that of shear (shear plane, dashed) and
departs from the finite deformation frame (straight line : trace
of foliation : dots : trace of lineation X).   Slip systems in
olivine : [100] (010) and in orthopyroxene : [001] (100). b) Illustration in the case of a natural peridotite, dextral shear. Equal
area projection, lower hemisphere ; contours: 1, 2, 4, 8 % for 0.45
% of the net area.   Solid triangle, best computed axis ; open
triangle, pole of best computed plane. 100 measurements for
olivine and pyroxene (a) Darot and Boudier, 1975. Petrology, 1,
225 ; b) Boudier, 1976. Thesis Nantes).

7.5.4. Relationships between stresses and preferred orientations

The preceding analysis shows that in the course of plastic deformation a crystal reorientates itself in response to the **geometrical solicitation** of its neighbours. If the aggregate effectively obeys the applied stress, each crystal obeys it only indirectly. This has two important consequences :
1) The angular variation between local stress (crystal) and applied stress (aggregate) will be much larger when the deformation is heterogeneous and,
2) heterogeneous deformation being the rule in rocks, theories that therefore seek to directly link the crystal orientation to the applied stress rest upon a doubtful foundation.

These reservations as to the direct relationship between the orientation of a crystal of the aggregate and the exterior applied stress do not prohibit in certain favourable cases, finding the stress directions and even their relative intensities, by a statistical treatment of measurements carried out on the deformed crystals. The method is similar to that used in stress analysis starting with measurement of a population of faults. We shall describe this method for limestones and marbles where one can profitably use the twinning of calcite but potentially it could apply to others rocks.

In response to an applied stress, a crystal of calcite twins or not depending upon its crystallographic orientation in relation to the orientation of this stress. In effect a threshold of shear stress must be passed in order to twin ; this threshold itself depends upon the angle of the twin plane with respect to the applied stress orientation (fig.2.15).

Only the crystals which are orientated in a favourable angular sector with respect to the applied stress orientation are twinned. From a study of the preferred orientation of the twinned crystals in a marble, one can deduce the orientation of the principal applied stress, either by a numerical inversion or by a

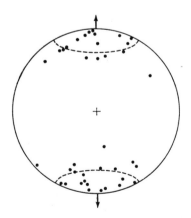

Fig.7.19. Dynamic analysis of an experimentally deformed marble. The orientations of the c axes of twinned calcite crystals are grouped in axial symmetry within a small circle centered on the N-S direction of tensional stress $\sigma 3$. (Lambert projection in a plane parallel to the axis of the test cylinder) (Turner and Weiss, 1963. McGraw Hill, New York).

graphical method (fig.7.19). The latter method also gives information on the relative intensity of the principal stresses
     In order to be successful applied, this method requires that
the crystals in the rock under consideration have a random orientation (absence of notable plastic deformation) and that the
deformation is weak.

FOR FURTHER READING

Bard, J.P., 1986. Microtextures of igneous and metamorphic rocks, D. Reidel,
Dordrecht, 278 pp.

Bouchez, J.L., Lister, G.S. et Nicolas, A., 1983. Fabric asymmetry and
shear sense in movement zones. Geol. Rund., 72, 401-419.

Gagny, Cl., 1983. Organisateur. Pétrologie structurale des roches érup-
tives. Bull. Soc. Geol. France, XXV, 3.

Marre, J., 1982. Méthodes d'analyse structurale des granitoides.
B.R.G.M., Manuels et Méthodes, 3, 128 p.

Vialon, P., 1979. Les déformations continues-discontinues des roches
anisotropes. Eclog. Geol. Helv., 72, 531-549.

Chapter 8

Continuous Heterogeneous Deformation;
Typical Structures

8.1. INTRODUCTION

In the field of continuous deformation, so far we have only considered homogeneous structures. The present chapter is therefore dedicated to an examination of the principal elementary structures of continuous heterogeneous deformation. We shall study structures that are only visible at the scale of a mineral grain such as pressure shadows or snowball inclusions, as well as structures visible at various scales such as shear zones or boudinage. Because of their importance in natural deformation and the voluminous knowledge acquired about them, folds are made the subject of a separate chapter although they are also included in the category of continuous heterogeneous deformation structures.

Some structures such as pressure shadows or tension fractures have already been mentioned in earlier chapters because their occurrence did not alter the homogeneous character of the deformation on the scale considered. Thus, heterogeneity introduced by pressure shadows around a porphyroclast at the scale of an aggregate disappears at the scale of the cleavage or foliation into which it is integrated. Here again we meet the concept of penetrative and non-penetrative deformation (§2.1.1.). Similarly, a shear zone, taken altogether, is a heterogeneous domain, although in the central part of such a zone the deformation may be considered as sufficiently homogeneous (fig.6.8b) to permit a kinematic analysis as described in the previous chapter. Thus one can resort to these methods of kinematic and dynamic analysis strictly valid only in the case of homogeneous deformation, if by changing scale, one can define new domains where strain is homogeneous. This operation is carried out usually by dividing a heterogeneous structure into sub-domains which can be considered as homogeneous ones (fig.8.12). Or inversely, it can be carried out by integration at a larger scale, as in the above example of pressure shadows, where the deformation is heterogeneous at the scale of a porphyroclast and becomes homogeneous at the specimen scale. This is also the case in deformation of orthogneiss, which is heterogeneous at the thin section scale (fig.7.3b and 8.3) and homogeneous on a field scale.

First we shall describe and analyse some heterogeneous deformation microstructures created by the presence, within a flowing matrix, of a rigid object which is generally a mineral which undergoes little deformation. We shall consider afterwards the most typical structures of heterogeneous strain, observable at all scales. These are shear zones, boudinage and folds (ch.9) which are the instabilities that correspond respectively to the three principal deformation modes : shearing, stretching and shortening.

8.2. MICROSTRUCTURES

The commonest microstructures are crystallizations in sheltered zones, more specifically pressure fringes and pressure
shadows, snowball and sigmoidal inclusions and fibrous crystal
growths in microfractures. In these three cases, the crystalline
growth that accompanies deformation constitutes a record of the
progressive deformation and can be used for kinematic purposes.
The nature of the minerals that crystallize during this deformation may be also symptomatic of a given metamorphism. Analysis of
this metamorphism and dating of the minerals allows the physical
conditions and age of the deformation to be determined.

8.2.1. Crystallization in pressure fringes and pressure shadows

Objects or hard crystals (pyrite, garnet, feldspar) that
are held in a more deformable matrix often have a double wake
along the X direction and possibly along Y which are called
sheltered zones (fig.8.1). The two essential points are the
nature of the minerals present in these sheltered zones and the
internal structure and relationship of these zones with the
surrounding cleavage or foliation. One speaks more readily of
pressure fringes when the infilling is composed of new minerals ;
the sheltered zone has a fibrous or lamella structure which is
independent of the surrounding cleavage.**Pressure shadows** relate
to the other cases, in particular when the infilling is composed
of the recrystallized parts of an augen.

In a medium where pressure-solution operates, the principal
minerals formed in pressure fringes are quartz, calcite and the
phyllosilicates, principally chlorite. Quartz and calcite grow in
fibres perpendicular to the free surface and chlorite in leaf-
lets parallel to it (fig.8.2). The free surface is opened during
deformation by a tensional crack which is generally located at
the contact with the hard crystal, in which case the most recent

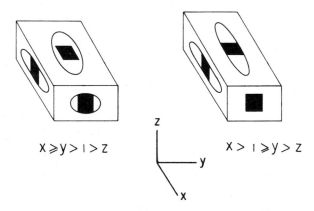

$x \geqslant y > i > z$ $x > i \geqslant y > z$

Fig.8.1. Sheltered zones around a rigid inclusion (black). On the
left, a dominantly flattening state and to the right, a constrictive one (Choukroune, 1971 Bull. Soc. géol. France, 257).

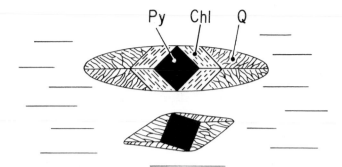

Fig.8.2. Pressure fringes with infillings of fibrous quartz (Q) perpendicular to the faces of pyrite (P) and chlorite (Chl) in lamellae parallel to them ; fracturing at the contact with pyrite. The fibres stay parallel during growth, suggesting a coaxial regime.

crystallization in a pressure fringe is the most internal. Cracks can open however at the external limit of the pressure fringe. Figure 8.2 illustrates the first situation where the infillings of chlorite and of quartz are orientated in relation to the central crystal and not to the limit of the fringe, which would be the case if the opening had occurred along this limit.

In rocks such as plastically deformed orthogneiss, the porphyroclasts of feldspars are also elongated by pressure shadows around which the foliation wraps, producing an augen structure in these gneisses. The infilling is commonly made up of recrystallization products of the porphyroclast (fig.8.3).

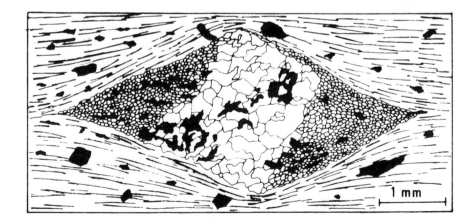

Fig.8.3. Recrystallization in pressure shadows of a feldspar porphyroclast in an augen gneiss. In contrast to pressure fringes the infilling consists of the same mineral as the hard object and is plastically deformed (Debat, 1974. Thesis Toulouse).

     The   minerals crystallized in the pressure fringe of   figure
8.2   show no curvature. The symmetrical character (upper figure)
or asymmetrical (lower figure) character of the infilling   simply
reflects   the   orientation originally adopted by the crystal   and
consequently has no particular significance. On the other hand as
shown in figure 8.4, the infillings are frequently curved without
the surrounding schistose fabric being deformed, which allows one
to   rule-out   the hypothesis of deformation   after   the   pressure
fringe formation.
     The   study of pressure fringes can assist the   determination
of   the strain regime and in the case of deformation due to shearing,   to   define the shear sense and possibly the   shear   strain
(fig.8.4b and c). Although it may appear simple, this analysis is
tricky   in   some   situations. Two theoretical points   may   guide
interpretation in complex cases.
     In   a progressive deformation,   the instant stretching of   a
pressure   fringe zone is carried out parallel to the X   direction
(and   possibly Y) of the infinitesimal strain   ellipsoid   ;   this
direction   is   itself at 45° to the shear plane in   simple   shear
and, in the pressure-solution case, is generally perpendicular to
the   principal stress $\sigma 1$. The fibres and lamellae which are growing during this increment of deformation may be not parallel   to
this direction as their growth direction is controlled by that of
the fracturing (see above and fig.8.2). The analysis must then be
carried   out   mainly on the shape of the pressure   fringes   themselves.   The   particular orientations of the crystallization products   allow us to know whether the hard object has been   rotated
and whether growth has come from its surface or from the tail   of
the   pressure   fringe. One immediate conclusion arises from   the
first point : in coaxial deformation, the pressure fringes tend
to be straight and the fibres, parallel (fig.8.2).
     The   second point deals with rotation of the   infillings   in
the course of non-coaxial deformation. The principle described in
   7.3   applies   here   also as there is no   assumption   about   the
nature   of the deformable medium. It applies firstly to the case
of   a   rigid   object   in   a   shearing   medium   which   undergoes   a
clockwise rotation for a dextral shear and a counterclockwise one
for   a sinistral shear (fig.7.8 and 8.4).   It applies also to the
pressure   fringe itself,   which if it is composed of minerals   as
ductile as a matrix, passively follows the rotation of the   axes
of the strain ellipsoid. Nevertheless, the bending takes place in
one sense or the other,   according to whether the initial opening
and   in consequence the growth of the sheltered zone operate from
the   rigid   object   (fig.8.4 and 8.5a) or from the   tail   of   the
pressure fringe (fig.8.5b).   The crack location depends upon   the
relative ductility of the hard object,   the matrix and the infilling minerals.   If the pressure fringe is formed of minerals that
are more rigid than the matrix such as fibrous quartz in a quartz
micaceous matrix,   or a quartzo-feldpathic aggregate around feldspar augen (fig.8.3), it tends to behave as a rigid object and to
undergo   rotation (fig.8.7).   This can result in a complex global
asymmetry.   In this case,   the infilling minerals are   themselves
deformed.

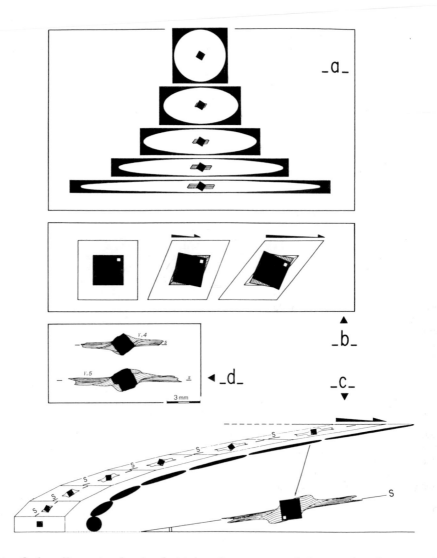

Fig.8.4. Numerical simulation of pressure fringes development by progressive strain and comparison with the natural example (d) of a pyrite crystal surrounded by quartz assumed to have no viscosity contrast with the matrix. a) Coaxial progressive pure shear with 30 % shortening increments. b), c) and d) Dextral simple shear; b) model of pyrite rotation and fringes growth ; c) results for shear strain Υ from 0 to 6 and detail for Υ =6 (Etchecopar and Malavieille. Jour. Str. Geol., in press).

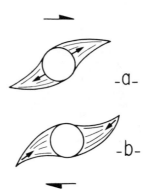

Fig.8.5. Asymmetry and curved infillings in pressure fringes in relation to a progressive deformation by shearing. The instantaneous growth in the sheltered zones takes place at 45° to the shear direction. a) growth from the surface of the hard crystal and b) growth from the ends of the pressure fringes.

8.2.2. Minerals with snowball and sigmoidal inclusions

Large minerals grown during metamorphism, the porphyroblasts, contain inclusions arranged in microfolds or spirals. Garnet, albite, andalusite and staurolite porphyroblasts are the commonest with inclusions. The inclusions are varied in nature ; quartz, amphibole, mica, graphite, magnetite etc. Two situations can be distinguished, depending on whether the porphyroblast develops during (**synkinematic or syntectonic growth**) or after (**postkinematic growth**) the deformation.

Where the crystallization of the porphyroblasts is syntectonic the inclusions form a double spiral : the **snowball or helicitic inclusions** as shown in figure 8.6. The double spiral results from the rotation of the porphyroblast whilst it is growing. The axis of rotation theoretically being the Y axis of the strain ellipsoid, analysis of such structures should be carried out in the X, Z plane. When the rotation of the inclusions measured on one arm of the spiral exceeds 90°, it can be stated that the deformation regime is one of shearing. The porphyroblast is at the same time a cause of instability due to the heterogeneity it induces, and a record of it. The sense of its rotation is unambiguously related to the shear sense. It is clockwise for a dextral shear (fig.8.6) and counterclockwise for a sinistral shear. Rotation, although considerable at times since angles exceeding 500 have been measured (Appendix II), constitutes only a minimal record of shear strain. The porphyroblast can cause recrystallizations in pressure shadows which it will tend to enclose during its growth (fig.8.7a and b). This explains the cavernous appearance of some garnets and the predominance of quartz inclusions (fig.8.7c).

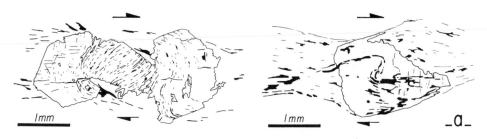

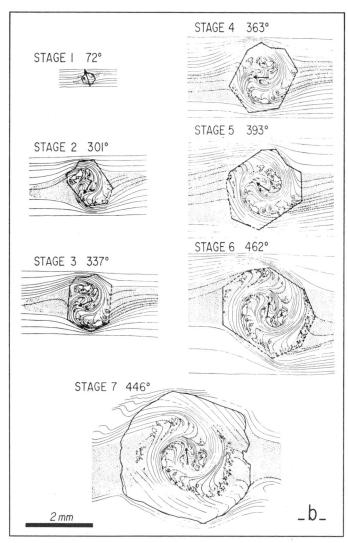

Fig.8.6. a) Porphyroblasts of garnet (left) and staurolite (right) with snowball inclusions. b) Reconstructed growth and rotation history of a garnet ; clockwise rotation associated with a dextral shear. (a) Pecher, 1978. Thesis Grenoble ; b) Powell and Vernon, 1979. Tectonophysics, 54, 25).

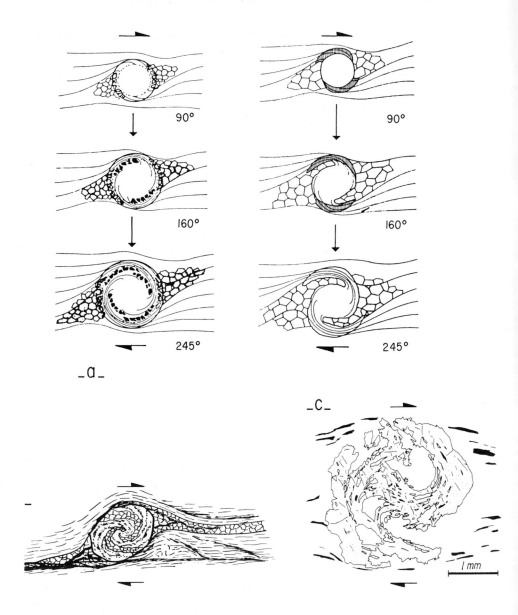

Fig.8.7. Spiral quartz-rich inclusions of pressure shadows formed at the time of growth of a garnet porphyroblast. a) Model
showing the mechanism of inclusion of the pressure shadows where
the growth rate of garnet with respect to the rotation rate is
high. To the left the growth of the garnet takes place at interfaces with quartz (in grey and then in black when it is included); on the right it is produced at interfaces with micas
(hatches). b) and c) Natural examples with in b) development of
pressure shadows. (a) Schoneveld, 1977. Tectonophysics, 39, 453 ;
b) Brunel, 1983. Thesis Paris ; c) Pecher, 1978. Thesis
Grenoble).

Porphyroblasts that develop after deformation can enclose **sigmoidal inclusions** when the fabric in which they grow is micro-folded (fig.8.8). It is as well to know how to distinguish these porphyroblasts with sigmoidal inclusions from porphyroblasts with snowball inclusions which, alone, record progressive deformation. The examination of a population of porphyroblasts shows a random character of the inclusion pattern in the first case and a relatively ordered one in the second.

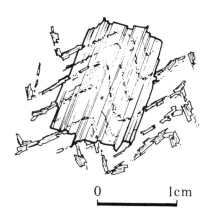

0   1cm

Fig.8.8. Sigmoidal inclusions in a porphyroblast whose growth is post kinematic (Bard, 1980. Masson, Paris).

8.2.3. **Mineral growth in tension fractures**

In hydraulic or fluid-assisted fracturing, both the tension fractures kept open by the fluid pressure and the broken-off lenses in faults (fig.5.15) tend to be sealed by minerals deposited by the fluids. These minerals reflect the prevailing physical conditions, for example quartz and/or calcite in rocks deformed at low temperature, quartz-glaucophane-epidote in metamorphism of blueschist facies or anorthite-pyroxene-olivine in a partly melted peridotite.

Mineral growth generally occurs as fibres or as parallel sheets. This growth is orientated and creates shape preferred orientations, and also lattice ones if the growth rate is anisotropic (fig.8.9). In the case of penetrative fractures, such orientations contribute to the development of a cleavage (fig.4.24). Besides this, the orientation of the fibres can represent a record of the relative displacement of the sides of the fracture, and in this way can provide a means of analysing the progressive deformation path.

However, trying to analyse the relationship between fibres and progressive deformation is beyong the scope of this book because of the complexity of the problem (fig 8.10 and 8.24). This problem is connected with that of pressure fringes (§ 8.2.1) in the sense that the orientation of the mineral fibres depend equally on several factors : location of the zone of growth which occurs at the fibres-fluid interface, either along one or both

sides of the fracture or within the fracture itself and following
a surface to be identified ; nature of the deposited minerals ;
effects of a possible subsequent plastic deformation affecting
the infillings ; fracturing mechanism which may operate   either
progressively and continuously, or discontinuously and repe-
titively by "crack-seal" (§4.3.1) ; lastly and above all,   mode
of growth of the fibres which may be either perpendicular at the
fibres-fluid interface, or parallel to the direction of relative
displacement of the sides of the fracture.

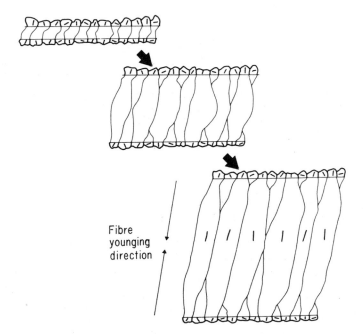

Fig.8.9. Development of shape and lattice preferred orientations
(the mark represents the crystallographic orientation of rapid
growth) during the opening of a crack-seal fracture. The fibres
are of the same mineral species as the minerals of the walls and
develop away from a median fracture (centripetal growth of fibres
is indicated by arrows). The competition between the numerous
nuclei progressively turns to the advantage of those crystals
where the direction of rapid growth coincides with that of the
opening of the fracture (Cox and Etheridge, 1983. Tectonophysics,
92, 147).

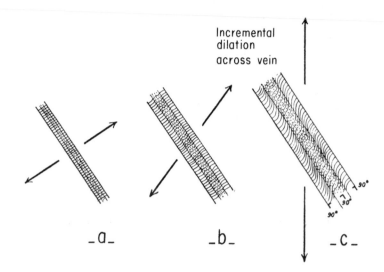

Incremental
dilation
across vein

_a_   _b_   _c_

Fig.8.10. Example of complex infilling in a tension fracture. a)
Progressive coaxial opening in the middle of the crack with
centripetal growth of fibres of the same mineral as in the walls.
b) and c) Opening becoming progressively oblique. Here the fibres
tend to follow the direction of opening in their growth. Following that their growth is centripetal (**syntaxial growth**) or
centrifugal (**antiaxial growth**), the bending is in "S" or "Z"
(Durney and Ramsay, 1973. Gravity and Tectonics, Wiley-
Interscience, New York).

8.3. SHEAR ZONE

8.3.1. Geometric and kinematic analysis

Shear zones have a characteristic structure (fig.6.8b and
8.11). In initially isotropic material, they are indicated by the
development of a very strong foliation which is attenuated as it
curves from one side to the other of the median band.
Geometrically there is an analogy with the rotation of a cleavage
from one part to another of a zone of intense solution (fig.6.4).
However, in the latter case the central zone shows no particular
trace of deformation and is typified by a abnormal concentration
of insoluble minerals.
In the most general case one can say that a shear zone
corresponds to a simple heterogeneous shearing with variable and
progressive deformation. The foliation which corresponds to the
path of the (X,Y) plane of the strain ellipsoid, gradually approaches the shear plane with increasing shear (fig.8.12) and the
stretching lineation, the shear direction. The shear sense is
found from the sense of foliation rotation from one part to the
other of the median band (fig.8.11). When the induced foliation
can be observed clearly, kinematic analysis can be carried out,
and when the entire shear zone structure is present the shear

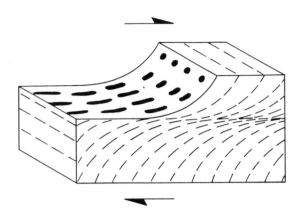

Fig.8.11.   Shear   zone with stretching lineation in the plane   of
the foliation.

strain can be measured across the shear zone, from which one
deduces the relative displacement of the two sides. Strain analysis is dealt with in Appendix II. We simply note here that the
relative displacement can be considerable. The value of γ' seems
to range between 10 and 100 and thus the displacement, between 10
and 100 km for a shear zone 1 km wide.

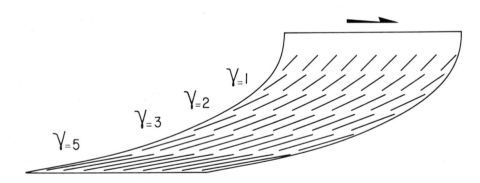

Fig.8.12.   Traces of the X axis (direction and length) of the
strain ellipsoid and strain γ in the vicinity of a shear zone
whose plane is horizontal.

Some typical structures of shear zones are secondary shears
and dragfolds. The S foliation formed on each side of the median
zone is frequently cut by a new C foliation which is spaced out
and discontinuous (fig.8.13). The superimposition of the S and C
surfaces gives the rock a spindle-shaped or augen appearance
(fig.8.14). The orientation of the C surfaces is close to the
general shear plane. Their identification thus constitutes a good
kinematic criterion (fig.8.18b).

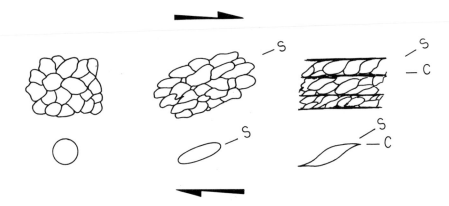

Fig.8.13.Diagram showing the formation of C ("cisaillement"=shear) surfaces and their relationship to the schistosity S (inspired by Berthé et al., 1979. Bull. Minéral., 102, 265).

In the center of the shear zone where strain is particularly large, one can often see folds in different stages of evolution. In the early stages, overturning is clearly visible in the section perpendicular to the shear plane and parallel to the lineation ; it can be used to determine the sense of movement (fig.8.14b). The continuing deformation separates the fold hinges within the deformed fabric and starts to rotate the axes in the direction of stretching. The folds then acquire a sheath-like shape (§9.5.2).

The relative displacements recorded in some shear zones pose the problem of where they end, both frontally and laterally. This problem has already been met within the termination of faults (§5.5.6). The model in figure 8.15a shows that in the ends of a movement zone, the shearing dies out in a relatively vast domain, respectively under compression and extension parallel to the movement direction. This theoretical scheme seems to be confirmed by natural observations (fig.8.15b). The model in figure 8.15a suggests that in a regime of simple shear, a situation that seems to be quite common in shear zones, and in the absence of deflections at their extremities, a dextral shear zone tends to curve clockwise and inversely, a sinistral shear zone, counterclockwise. Note that the appearance of Riedel faults in the zone of propagation of a fault causes the same rotation (§5.5.6). This rotation would explain the coalescence of shear zones of the same sense (fig.8.16). The dying-out of the relative displacement on each side of the shear zone, induced by this rotation also explains the splaying-out of some shear zones in horse-tails at their termination (fig.8.16b). In the case of relatively superficial thrusts in which the front rises up to emerge at the surface in reversed faults, this problem of termination does not arise (fig.8.17). In this figure, it can be also seen that laterally a thrust grades into a steep or vertical shear zone and upward into a strike-slip fault.

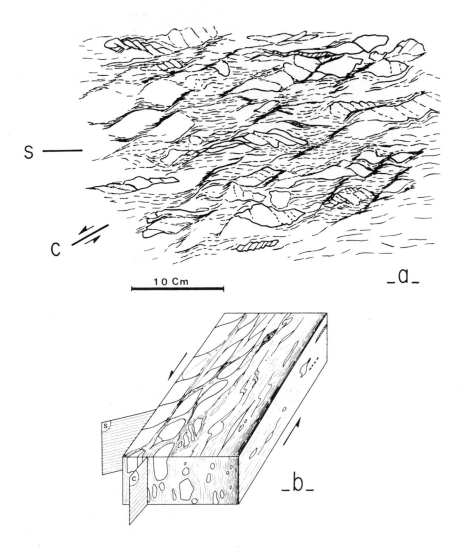

S ———

C

10 Cm

_a_

_b_

Fig.8.14. Planes of discontinuous shearing C and foliation planes
S. a) The relation between S/C in an augen gneiss defines the
shear sense, here sinistral. b) Transition between the centre and
border of a shear zone (from right to left). The C surfaces which
are parallel to the shear plane cut the schistosity S forming
asymmetrical augen. Also note the appearance of drag-folds. The
overturning sense and asymmetry of the augen indicate the shear
sense (a) Brunel, 1983. Thesis Paris ; b) Burg et al., 1981.
Tectonophysics, 78, 161).

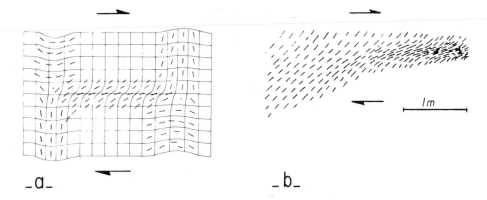

_a_   _b_

Fig.8.15.   Longitudinal dying out of shear zones (seen in the   XZ
plane of strain). a) Geometrical model in plane deformation ; the
lines correspond to the X axis of the strain ellipsoid.   b) natural   situation ;   the lines correspond to the foliation. (a)
Ramsay,   1980.   Jour.   Str. Geol., 2, 83 ; b) Ramsay and Allison,
1979. Bull. Suisse Min. Petr., 59, 251).

-a-

-b-

Fig.8.16.   Clockwise curvature associated with a   dextral   shear
zone,   which is   responsible for the dying out of the   relative
displacement. a) Coalescence of shear zones. b) Horse tail termination of a shear zone.

Figure 8.18 presents a synthesis of the principal shear
criteria usable at different scales in a shear zone.

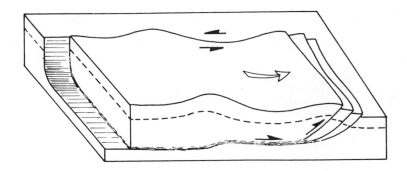

Fig.8.17. Relationships between a vertical shear zone and a deep thrust on the crustal scale (after Coward, 1980. Jour. Str. Geol., 2, 19).

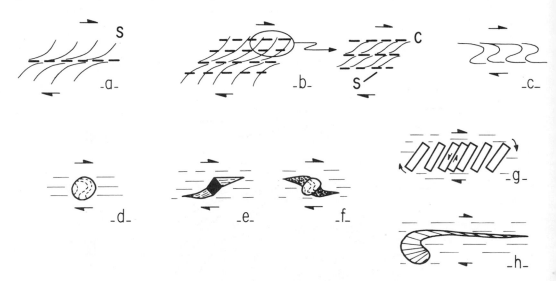

Fig.8.18. Shear criteria in heterogeneously deformed rocks. Fine lines : foliation ; thick dashes : shear plane. a) Sigmoidal foliation. b) Intersection of the S foliation with the disconti- nuous C shear planes : characteristic spindle or fish shape. c) Small drag folds. d) Porphyroblast with snowball inclusions. e) Crystallization in pressure fringes (fibrous or lamellar minerals that may be curved but undeformed). f) Pressure shadows (deformed minerals).g)Stretching with reversed shearing of a porphyroclast. h) Retort-shaped ("Cornue") porphyroclast (starting from an unfa- vourable orientation under shearing, the crystal undergoes ben- ding which enables the upper part to glide freely, see § 7.5.2. and fig.II.18).

8.3.2. **Microstructural analysis**

Shear zones typically develop in mechanically homogeneous metamorphosed rocks in anhydrous or slightly hydrated retrograde conditions, as in the case of a granite undergoing shear in amphibolite facies conditions (500°C). The shearing is then often plane (simple shear) and deformation mechanisms are plastic in constant volume conditions. Some natural situations deviate from this ideal case, if for example there is a flattening component with stretching in the Y direction, or volume variations related to fluid circulation and the intervention of pressure-solution processes.

The increasing intensity of ductile deformation from the edges towards the centre of the shear zone is expressed by the increasing stretching and recrystallization of the porphyro-clasts. The central zone is dominantly made up of minute neo-blasts (fig.8.19). These fine grained (a few tens of micrometres) and strongly deformed rocks are termed **mylonites**. If grain boundary sliding is facilitated by intergranular diffusion (relatively high temperature, presence of fluids), the very small grain

10 cm

Fig.8.19. Microstructure in the border zone of a sheared mica-ceous peridotite. The central mylonitic structure results from increasing recrystallization of porphyroclasts (Brodie, 1980. Jour. Str. Geol., 2, 265).

size favours a transition from plastic to superplastic flow
(§4.2.3). Plastic and superplastic behaviours are illustrated in
figure 8.20. The **structural softening** (reduction of grain size,
development of preferred orientation and crystallization of new
minerals) which takes place in the central parts of the shear
zones can concentrate the deformation to the point where it
creates a very sharp boundary between the shear zone and the
underformed surroundings (fig.8.21). The deformation within such
bands can be considerable, leading to rocks in which the grain
size does not exceed a few microns and which are called **ultramylonites**. The deformation mechanism is often superplastic. At the
crustal scale, the structural softening explains why old shear
zones are reactivated again and again. Thus a shear zone created
by crustal extension during rifting, can serve as a thrust plane
during a subsequent crustal collision.

Softening in a shear zone can have other origins than the
structural one just considered. It may be due to chemical causes
(alteration by a fluid phase § 3.3.2) or to physical causes,
thermal variations, for instance caused by **shear heating.** In this
latter case, the movement dissipates energy in the form of heat,
which yields Q(t) per unit time, where Q(t) = $\dot{\varepsilon}$ xσ, with $\dot{\varepsilon}$ being

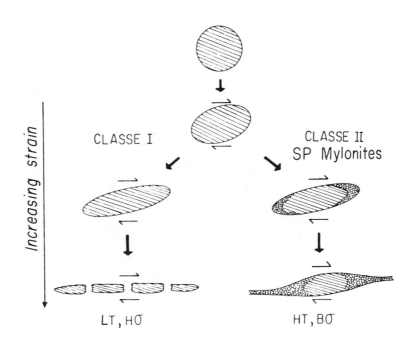

Fig.8.20. A model of the development of mylonites in shear zones.
The increasing deformation can lead, in the plastic domain (class
I, low temperature, high stress), to an increasing stretching
with limited recrystallization of porphyroclasts, or in the
superplastic domain (class II), to an intense recrystallization
favouring grain boundary sliding and the redistribution of porphyroclasts as a micro-banding (Boullier and Gueguen, 1975.
Contr. Min. Petrol., 50, 93).

the strain rate and σ , the deviator. The thermal conductivity of
rocks being low, the shear heating raises the temperature of the
zone by tens to hundreds of degrees which has an important effect
on its ductility (§4.2.3). Fusion can be attained if, as in a
seismic fault, the movement is sufficiently rapid to prevent heat
dissipation. This is shown by the injection and brecciation in
the centre of the shear zones by blackish veins called **pseudotachylites**. These have a structure comparable to that of some
devitrified lavas with a minute grain size (a few microns) and
fluidal bands enclosing mylonitic fragments. Figure 8.21 shows
that the movement responsible for this fusion is produced within
ultramylonitic bands.

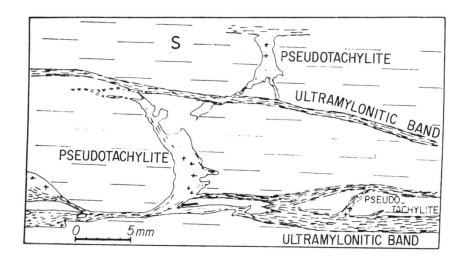

Fig.8.21. Relationship between ultramylonitic and pseudo-
tachylitic bands in a mylonitic zone (Passchier, 1982. GUA Pap.
Geol., 16).

8.4. STRETCH STRUCTURES ; BOUDINAGE

The study of normal faults has shown the importance of
tensional structures in crustal deformation, and also underlines
the relationship between superficial brittle deformation and
deep-seated ductile deformation (§5.5.1). We now examine the
principal manifestations of this tensional ductile deformation.
In formations which contain inclusions or beds that are more
competent than their matrix, stretching leads to **boudinage**, that
is to say, the segmentation of the more competent material into
fragments, the **boudins** which are elongated and aligned in the
(X,Z) plane of the deformation in the manner of a string of
sausages (fig.8.22, 6.15c, 6.16c). The stretched zone between the
competent boudins is infilled by the flow of the matrix and/or by
the deposition of new minerals indicating an early stage of
extensional fracture (fig.8.23). The fractures are initiated
directly, or after a stage of plastic stretching called **necking.**

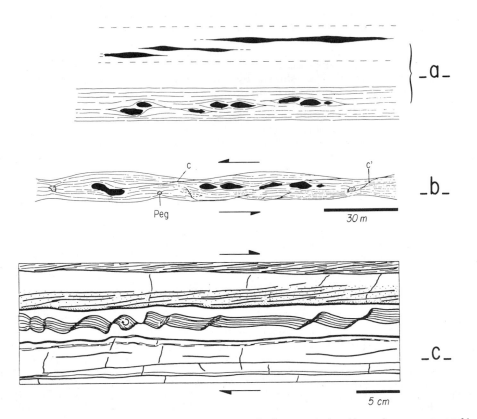

Fig.8.22. Boudinage of competent beds seen in the plane perpendicular to the foliation and parallel to the lineation. a) Essentially co-axial deformation ; the boudins are stretched in the
direction of alignment. b) Non-coaxial boudinage ; note the
stretching of the matrix along the C surfaces. c) Non-coaxial
boudinage with the boudins being stretched obliquely to their
direction of alignment. (a and b) Van der Driessch, 1983. Thesis
Montpellier) ; c) Malavieille, pers.com.).

In a zone of necking, the sectional surface of the neck upon
which a supposedly constant tensile force operates diminishes
progressively ; hence the stress increases and passes above the
limit of ductility of the bed, causing rupture (fig.2.10).
    In the (X,Y) plane of the cleavage or foliation, the boudins
are roughly elongated in the Y direction, that is to say that
their planes of segmentation are at high angles to the stretching
direction X (fig.6.16c). They may also be stretched along Y and
thus have a subordinate direction of fracturing perpendicular to
this new axis. The layer then breaks up into prisms elongated in
the Y direction. Stretching in plane deformation (Y invariant)
may also lead to this boudinage by fracturing in two directions
because stretching along X commonly induces two families of
fracture planes between 45° and 90° on each side of X (fig.8.24).

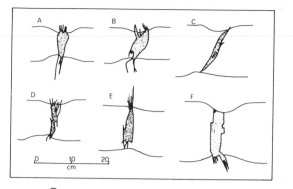

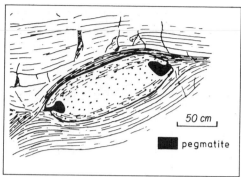

_a_ _b_

Fig.8.23. Necking of layers and infillings of tension fractures.
a) Boudinage of calcareous beds with infillings of quartz and
calcite. b) Amphibolites in quartzites. (a) Beach and Jack, 1982.
Tectonophysics, 81, 67 ; b) Lloyd and Ferguson, 1981. Jour. Str.
Geol., 3, 117).

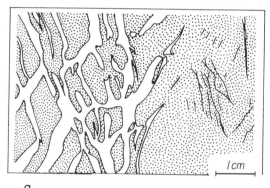

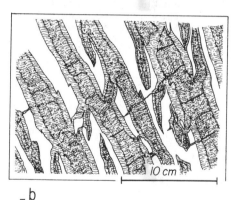

_a_ _b

Fig.8.24. Boudinage in "chocolate tablets" (seen in the foliation
plane) due to the segmentation of a competent layer along two
sets of planes at a high angle to the tensional X axis (E-W). a)
Conjugate fracture planes at 65 (principal plane) and 90°
(subordinate plane) to X in a limestone bed in a sequence of
marls. The infilling by quartz fibres (dotted) and calcite has a
complex history. (a) Burg and Harris, 1982. Tectonophysics, 83,
347 ; b) Casey, Dietrich and Ramsay, 1983. Tectonophysics, 92,
211).

By analogy with **Lüders bands** described in metals, it has been
proposed that necking, which initiates the boudins, occurs following the invariant directions of the (X,Y) ellipse of deformation
(see Appendix II). The contrast in ductility between the boudin
and its matrix controls whether ductile deformation precedes
rupture or not. When this contrast is weak, rupture does not take
place but a simple succession of nodes and bulges along the less
ductile bed.

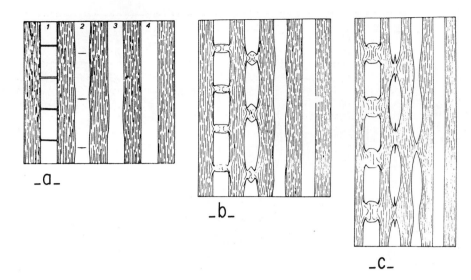

Fig.8.25. Relations between the stretching of beds and their ductility contrast to the matrix, under increasing deformation from left to right. The contrast in ductility decreases from bed 1 to bed 4, the latter having the same ductility as the matrix. Observation in the (X,Z) plane of deformation. (Ramsay, 1967. McGraw Hill, New York).

Ductile extension structures in a foliated homogeneous medium are less spectacular. They appear to follow local concentrations of deformation, for example linked to a structural softening (§8.3.2) in local shear zones (fig.8.26a) or to hydraulic fracturing at a high angle to the foliation (fig.8.26b).

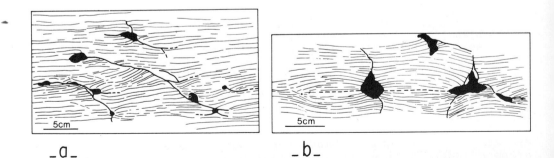

Fig.8.26. Tension structures in a homogeneous foliated medium (quartz fracture infillings in black). a) Dextral shear passing into boudinage with fracturing, due to non-coaxial extension. b) Boudinage initiated at extension fractures due to coaxial extension (Platt et Vissers, 1980. Jour. Struct. Geol., 2, 397).

     In the case of progressive coaxial stretching, the elonga-
tion of the boudins is parallel to their alignment in the foliation (fig.8.22a) and in a homogeneous layered medium, if the
stretching occurs in shear zones, these form a conjugate system
related to the layering (fig.8.27). In the case of a progressive
non-coaxial stretching of boudins, their elongation is oblique to
their alignment, forming step-like offsets (fig.8.22b and 8.18g)
and the shear zones form a single system. The non-coaxial regime
can be due to shearing on the C surfaces described previously ;
the relative orientations of S and C allow the shear sense to be
found (fig.8.18b).

Fig.8.27. Model of coaxial extension caused by conjugate shear
zones (Ramsay, 1980. Jour. Struct. Geol., 2, 83).

FOR FURTHER READING

Ramsay, J.G. et Graham, R.H., 1970. Strain variation in shear belts.
Canadian Jour. Earth Sc., 7, 786.

Spry, A., 1969. Metamorphic textures. Pergamon ed., Oxford, 350 p.

Chapter 9

Folds

9.1. INTRODUCTION

Folds are expressed by the warping of a reference surface. They belong to the category of continuous heterogeneous deformations and are in fact their most spectacular example . The reference to a surface is important in two respects : it is thanks to the deformation of reference surfaces that we can observe folding in sedimentary and metamorphic rocks, and secondly, the bedded or stratified nature of these rocks favours the generation of the instabilities that produce the majority of folds. It is important to be aware that folding can affect other rock types. Magmatic flow and plastic deformation of homogeneous formations can also be accompanied by folding, which is clearly visible when the medium possesses a definite layering (fig.7.2 or 7.4b), but more difficult to identify when it only has a lamination or foliation plane (fig.7.3a).

Until recently, the fascination of folds and their ubiquity have made them the main subject of research in structural geology, which is disproportionate to their contribution to finite strain of tectonites. Essential in superficial domains, where folding of sedimentary formations can reach a few tens of per cent in shortening, their contribution becomes relatively negligible in large scale ductile deformation. The folds only then represent the results of generally passive transposition of an initially localised heterogeneity in a globally homogeneous deformation. The classic studies of folding have been made in superficial domains where the fold axes are in general perpendicular to the principal X direction of the deformation and where the regime is habitually one of coaxial shortening and moderate straining. Used without caution on the structures of large ductile deformations, the results of the preceding studies have been given too much weight in the interpretation of these structures. Thus the importance of coaxial flattening has been exaggerated in relation to shearing. Similarly, stretching lineations have themselves been considered sometimes as orthogonal to the direction of ductile flow due to their parallelism with the fold axes (fig.6.17).

In view of the objectives aimed at here and the large literature on folding, this subject is dealt with relatively briefly.

9.2. GEOMETRICAL ANALYSIS OF FOLDS - THE CASE OF AN ISOLATED SURFACE

Before examining structures characteristic of the folding of

a bed of a given thickness, we shall define the geometrical
properties, firstly intrinsic, then related to a spatial
reference, in the case of an isolated surface. This theoretical
situation is approached in nature when a thin and competent layer
such as a quartz vein is enclosed in an incompetent matrix, such
as a mudstone.

9.2.1. Morphology of folds

The principal geometrical and morphological definitions of
folds are illustrated in figure 9.1 ; **hinge** = region of the
smallest radius of curvature ; **limbs** = region of the largest
radius of curvature ; **axial surface** = bisectrix of the angle
between the limbs. Note that in the case of a folded series the
axial surface is defined as passing through the successive fold
hinges (fig.9.17) ; **axis** = locus of the points of maximum curvature or intersection of the folded bed and its axial surface.

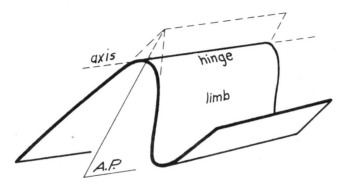

Fig.9.1. General geometrical features of a fold.

If the axial surface is straight, the fold is **plane or
planar** (fig.9.2a and b) ; it is non-planar in other cases
(fig.9.2c and d). If the axis is a straight line and the surface
of the fold can be generated by displacement of a line parallel
to the axis, it is called a **cylindrical fold** (fig.9.2a and c). It
is non-cylindrical in the other cases (fig.2.9b and d). A **conical
fold** is a non-cylindrical fold in which the surface can be reproduced by rotation of a line about a fixed point (fig.9.2b). A
sheath fold is a conical fold in which the conical surface is
completely closed in the neighbourhood of the cone's apex(fig.9.3
and 6.17b). The corresponding cross sections have a typical
rosette form (fig.9.3b). These various definitions only apply at
a given point in a fold in view of its longitudinal variations.
Thus in a cylindrical fold the ends are conical, and called
periclinal terminations in structural cartography.
These morphological descriptions must be complemented by
measurements of the amount of curvature in the hinges : angular
fold (fig.9.18), concentric (fig.9.11a), box-shaped (fig.9.11b),
etc. and of the angle of opening : open fold, tight or isoclinal
if the two limbs are practically parallel (fig.7.4b).

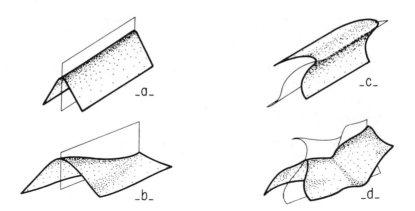

Fig.9 2. Shapes of the axial surface. a) Plane cylindrical fold.b) Plane and conical fold. c) Non-plane cylindrical fold.d) Non-plane, non-cylindrical fold.

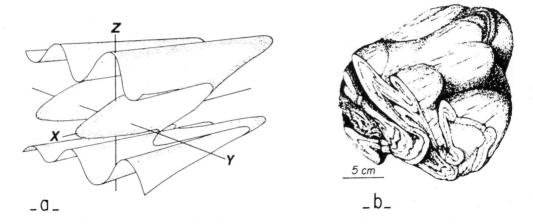

Fig.9.3. Sheath folds. a) Relationships to the principal direc-tions of the finite strain ; the axis of the fold coincides with X and the axial plane is perpendicular to Z. b) Natural example. (a) Henderson, 1981. Jour. Str. Geol , 3, 203 ; b) Faure and Malavieille, 1980. C.R.A.S., 290, 1349).

9.2.2. Orientation of folds

The orientation of a fold is defined with respect to the horizontal plane. It relies on the orientation of the axial surface (fig.9.4) and the axis (fig.9.5) i.e. **upright** or **normal** fold if the axial plane is vertical, **inclined** if the steeper (lower) limb does not pass the vertical and **overturned** if it does pass it, and lastly **recumbent** if the axial plane is horizontal. A fold is **horizontal, plunging** or **vertical** if its axis is horizon-tal, plunging or vertical.

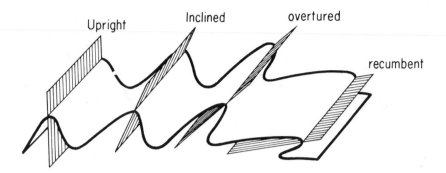

Fig.9.4. Definitions relating to the orientation of the axial surface of folds.

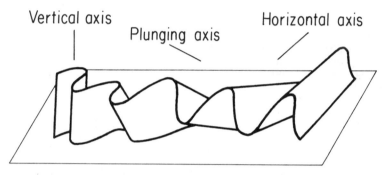

Fig.9.5. Definitions relating to the orientation of the fold axis.

9.2.3. Systems of folds

A fold is rarely isolated and often belongs to a system of more or less identical folds characterised by periodic repetition of simple folds which have a given symmetry (fig.9.6). When the folds are **asymmetric** this character is precised by the sense of overturning. The overturning is dextral or clockwise in figure 9.6b. In figure 9.7 the overturning is, from left to right, firstly dextral, then sinistral or counterclockwise and finally to the right again, dextral. This definition is ambiguous for if the observer turns around, the sense of overturning is reversed. In the case of plunging folds, this ambiguity is avoided by conventionally looking at them downward in their direction of plunge, and distinguishing Z (dextral) and S (sinistral) folds. Concerning the shape of the envelope surface of elementary folds, it can often be shown that it is folded in a comparable manner (fig.9.7). With the aid of this envelope one can define the **n order**, whereas simple folds, often called parasitic folds are affected by an order of n + 1. This idea of orders of folding is fundamental as it is this which justifies microtectonic study of folds. Thus the study of the style and spatial disposition of

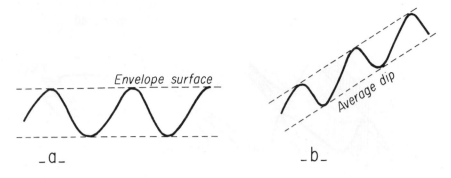

Fig.9.6.   Periodic fold systems. a) Symmetrical. b) Asymmetrical, dextral overturning. Envelope surfaces are dashed.

small folds that are visible at the outcrop scale can disclose the form of much larger folds (fig.9.8).
    Another system of folding consists of the association of **folds en échelon** (fig.9.9) so called because of the disposition of their axes. Such strongly non-cylindrical folds appear in the sedimentary cover overlying a shear zone affecting the basement.

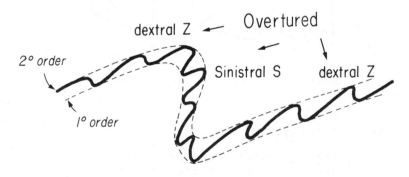

Fig.9.7.   System of second order folds (thick line) whose envelopes define a first order fold (dashes).   The overturning of the small folds varies from one limb to the other of the first order fold.

Fig.9.8.   Reconstruction of a first order fold (dashes) from outcrops showing second order folds.

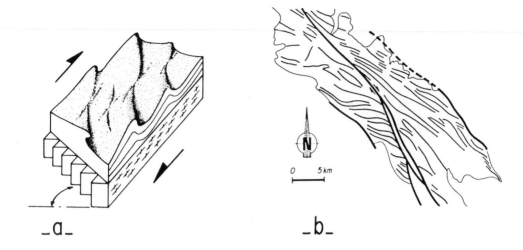

_a_   _b_

Fig.9.9. En échelon folds. a) Model of folding developed in the
cover formations over a sheared basement. b) Folds in Plio-
Pleistocene cover rocks along the San Andreas fault system
(Salinas region, California). (a) Iglesias Ponce de Leon and
Choukroune, 1980. Jour. Str. Geol., 2, 63 ; b) Page, 1981. Rubey
Vol.1, Prentice-Hall, N.J.).

9.3. FOLDING OF AN ISOLATED BED

The shape of folds, for a given bed thickness and folding
mechanism, depends largely on the contrast in viscosity between
the bed and its surroundings. We shall now describe two extreme
cases, that of parallel folds which corresponds to a large viscosity contrast, and that of similar folds which ideally corresponds to no viscosity contrast. Natural folds result very often
from situations intermediate between these two cases.

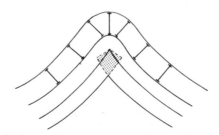

Fig.9.10. Model of a parallel fold. The segments perpendicular to
the bed surfaces are all the same length. The dotted area in the
core of the fold corresponds to the theoretical excess volume.

9.3.1. Parallel folds

Description

    Parallel folds are those in which the thickness of the layer, measured normal to the layer, remains constant everywhere (fig.9.10). Its radius of curvature decreases into the core of the fold and inversely, grows towards the exterior. Particular cases correspond to **concentric folds** in which the hinges have a circular arc in profile (fig.9.11a), to **box folds** which have two conjugate axial planes (fig.9.11b) and to **ptygmatic folds** that result from folding of a medium with a very large viscosity contrast with its matrix, for example a pegmatite vein in phyllites (fig.9.11c).

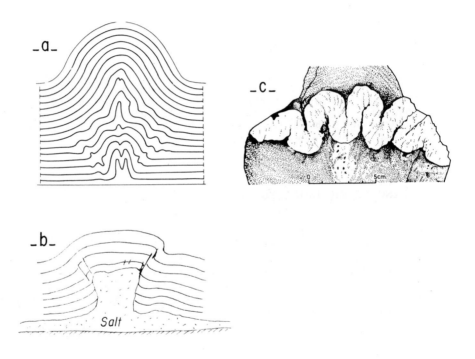

Fig.9.11. Types of parallel folds. a) Concentric parallel fold in the upper part and disharmonic folds in lower part. b) Box fold with two conjugate axial planes. c) Granitic ptygmatic fold in a gneiss. (a) Goguel, 1952. Masson, Paris ; c) Kuenen, 1968. Tectonophysics, 6, 143).

Mechanisms of formation

    Parallel folds can be formed either by buckling (fig.9.12a), flexural slip (fig.9.12b) or flexural flow (fig.9.12c). Folds produced by **buckling** tend to have the deformation concentrated in the hinge zone (**hinge deformation**). A surface of no finite strain, commonly referred to as the **neutral surface**, separates the outer arc of extensional deformation from the inner arc of

compressive deformation. The intensity of the extensional and
compressive deformation increase with fold tightening. In **flexural folding**, however, deformation is concentrated in the limbs of
the folds (**limbs deformation**), with the layering acting as a
shear plane that does not undergo deformation (invariant surface,
see § 2.4). In **flexural slip** (fig.9.12b) shearing takes place
preferentially along boundary layers, commonly in the more ductile intervening layers between the competent beds. **Flexural flow**
on the other hand is a distributed shear throughout the beds
(fig.9.12c). The shearing undergone by the intervening ductile
beds can induce **drag folds** (higher order folds) in which
asymmetry characteristises the sense of displacement (fig.9.13).

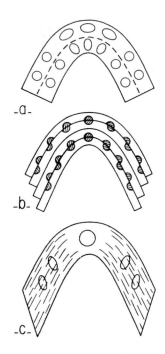

Fig.9.12. Mechanisms of folding. a) Buckling : the limbs are
virtually unaffected during the folding (circles), the deformation is concentrated in the hinge zone with E-W stretching on the
outside and N-S stretching on the inside ; neutral surface,
dashed. b) Flexural slip and c) flexural flow : the hinge zone is
almost unaffected (circles) and the limbs are sheared, b)discon-
tinuously in flexural slip, c) continuously in flexural flow.
(Modified from Arthaud, 1970. Publ.Univ.Sc.Tech. Languedoc).

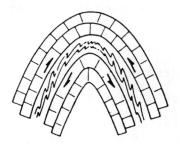

Fig.9.13. Drag folds formed in a ductile bed by the shearing due to flexural-slip of competent beds above and below it.

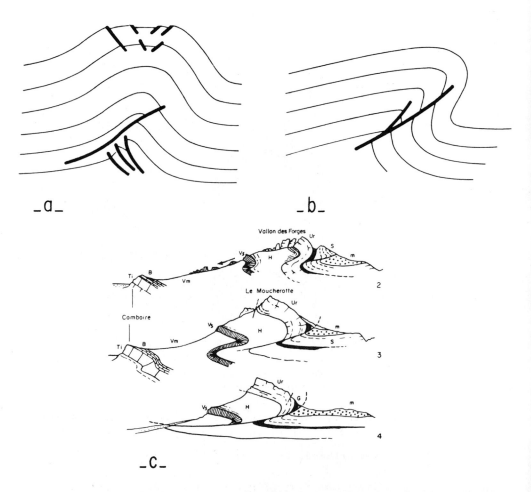

Fig.9.14. Discontinuous deformation associated with hinge deformation of parallel folds. a) Upright symmetrical fold with conjugate faults in the core and normal faults on the exterior. b) Overturned fold passing inwards into a faulted core. c) Evolution of an overturned fold into a small thrust (sub-Alpine chain of Vercors). (c) Debelmas, 1970. Masson, Paris).

Associated discontinuous deformation

When a thick and uniform bed is folded, the local stress concentration can exceed the plastic limit of the material, inducing rupture. In a hinge deformation the stress is tensional on the outside of the fold where it causes tension cracks that are close to being perpendicular to the beds (fig.5.1) and normal faults (fig.9.14a). Alternatively, it is compressional on the inside arc, causing reverse faults, conjugate (fig.9.14a) or not (fig.9.14b), depending on whether the fold is symmetrical or asymmetrical. Faults also develop on the limbs of the folds during limb deformation. These faults, which conform to the flexural slip mechanism, tend at first to utilise the weaker

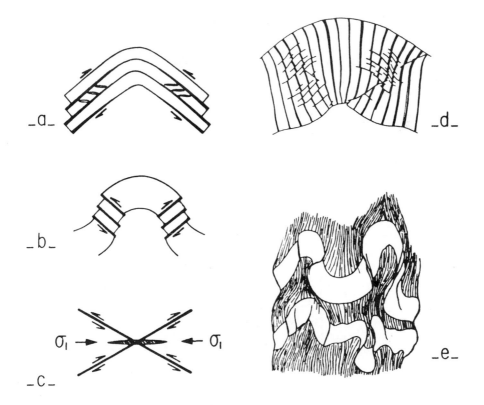

Fig.9.15. Discontinuous deformation associated with flexural slip in parallel folds. a) Faults (thick lines) and en échelon tension fractures caused by bed over bed slip. b) Conjugate faults along the lines of maximum strain. c) Relationship between the various directions of faulting and the principal stress direction $\sigma 1$, assumed to be normal to the axial plane of the folds. d) Experimental deformation of a homogeneous clay plate (the stripes are drawn to indicate the deformation) illustrating the appearance of a main fault system parallel to the edges of the plate and additionally, of reverse faults.e) Microfolds with limb deformation by reverse faulting in sandstone within a pelitic series. (d) Kuenen and De Sitter, 1938. Leidse Geol. Med., 10, 217 ; e) Arthaud, 1970. Publ. Univ. Sc. Tech. Languedoc).

horizons separating the beds (fig.9.15a and d), but a new system
cross-cutting the beds may develop along transverse directions
(fig.9.15b and e). The shear stress acting parallel to the beds
can also open en échelon tension cracks (§5.3).

Shortening perpendicular to the axial plane of parallel
folds suggests that the $\sigma 1$ stress is also perpendicular to this
plane. On the contrary, the $\sigma 2$ and $\sigma 3$ stresses do not need to be
respectively parallel to the fold axis and perpendicular to it in
the axial plane (§9.5.2). Fig.9.15c shows that the system of
faults and fractures caused by flexural folding is compatible
with the application of $\sigma 1$ stress perpendicular to the axial
plane of the folds.

Development of parallel folds

When a competent bed of thickness t, viscosity $\eta 1$, and
surrounded by an incompetent medium of viscosity $\eta 2$, is compres-
sed longitudinally it tends to develop folds in which the wave
length w (fig.9.16) is related as follows :

$$w = 2\pi t \sqrt[3]{\eta 1/6 \eta 2}$$

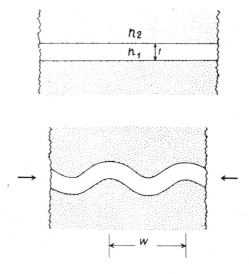

Fig.9.16. Relationships between the thickness t of a bed, the
viscosity contrast between beds and the wavelength w of folds
(Ramsay, 1967. McGraw Hill, N.Y.).

This relationship, established theoretically and verified
experimentally, applies in the case of rocks where the folding is
open. After measuring in the field w and t, one can deduce the
viscosity ratio assuming it to be Newtonian. If the contrast of
viscosity between the bed under consideration and the matrix is
weak, initiation of folding is difficult and the medium undergoes

a shortening that is very nearly homogeneous. If the contrast is large, the competent bed controls the deformation of the medium (fig.6.12). In a stratified medium it is the thickest bed which imposes the wavelength and gives the general form to the folding. The thinner beds then develop folds of smaller wavelength (folds of a higher order) of which the envelope surface follows the folds of the thick bed.

9.3.2. Similar folds

Description

Similar folds are those in which the two surfaces of the bed under consideration, when measured parallel to the axial plane, remain at all points a constant distance apart (fig.9.17). In other words, in a translation by this amount the two surfaces coincide ; geometrically, the folds affecting the two surfaces are similar. The formation of similar folds requires a movement of material within the bed from the limbs which become thinner, to the hinges which swell. This migration becomes all the more important as the fold tightens, so that in some isoclinal folds the stretching of the limbs leads to the separation of the bed into lenses enclosing the hinges (**intrafolial folds**, see fig.6.10). In the case of similar folds, internal flow of the bed is thus greater than in the case of parallel folds, thus explaining why in a stratified medium, the competent beds may form parallel folds and the incompetent beds, similar folds.

As examples of particular types we mention **chevron folds** (fig.9.18a) and **kinks** (fig.9.18b) ; these are folds with angular hinges, which are respectively symmetrical and asymmetrical.

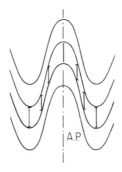

Fig.9.17. Model of similar folds. The segments shown parallel to the axial plane retain the same length at every point between the boundaries of the bed under consideration.

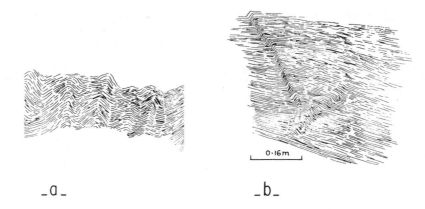

_a_   _b_

Fig.9.18. Angular similar folds. a) Chevron folds. b) Conjugate
kink folds (Williams, 1970. Tectonophysics, 10, 437).

Mechanisms of formation

In terms of folding mechanisms, one can oppose **buckling** and
flexuring described above and **simple heterogeneous shearing.** This
can be   shown   by the two ways of deforming a   pack of   playing
cards.   One can either bend it (fig.9.19a),   incidentally   with
flexural   slip,   or   cause gliding of some cards with respect   to
others (fig.9.19b) which is then   simple   heterogeneous   shear.
Natural folds derive their complexity from the fact that to these
modes of deformation, whether associated or not, often is added a
component   of coaxial flattening perpendicular to the axial plane
of the folds ; flattening may or may not be homogeneous. The case
of similar folds is an illustration of this complexity.
Only simple heterogeneous shear is capable of causing strictly similar folds. As shown ·in fig.9.20,   this shearing operates
parallel   to the axial plane of the fold.   The plane of   shearing
being invariant during folding,   the length of the segments   contained   in   this plane remains constant,   thus conforming to   the

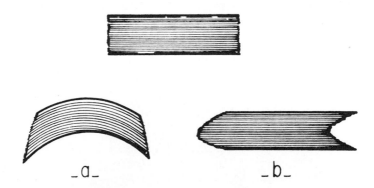

_a_   _b_

Fig.9.19. Contrasting folding mechanisms shown by the deformation
of a pack of cards. a) Flexural slip. b) Simple   heterogeneous
shear.

definition of similar folds. Shearing can be continuous or discontinuous ; in the second case it may be guided by an earlier cleavage. It then results in **microlithon folds** (fig.9.21, 6.3e and 6.5). If shearing is continuous, it necessarily entails a divergent fanning out of the cleavage from the hinges (fig.9.20 and 6.10a).

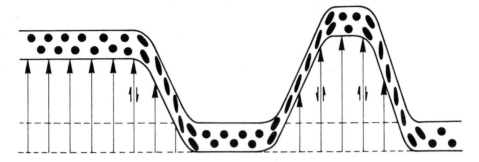

Fig.9.20. Formation of similar folds by simple heterogeneous shear (the arrows show the amount of displacement in an initially horizontal bed). The change of small circles into ellipses in the shear zones shows that the schistosity, which lies parallel to the long axes of the ellipses, forms a fan diverging from the hinges.

Fig.9.21. Microlithon fold in weakly consolidated volcanic tuffs. The fold is defined by the envelope of the beds sheared by reverse faults.

A **coaxial heterogeneous flattening** can cause in an adjacent domain the regime of simple heterogeneous shear necessary for the development of similar folds (fig.9.22). If the flattening is considerable, the folds that have developed in the zone of heterogeneous flattening themselves adopt a style that is very nearly similar. The longitudinal variations in the flattening strain curves the axis of the fold (fig.9.23).

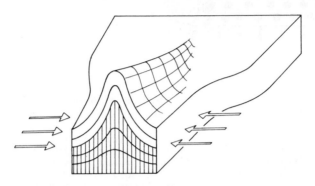

Fig.9.22. Formation of similar folds (beds in white) related to heterogeneous flattening (hachured beds). In the most deformed part of the hachured zone the folds are nearly similar.

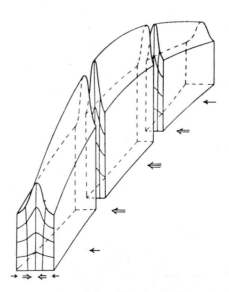

Fig.9.23. Curvature of a fold axis in relation to the longitudinal variation of compression (Ramsay, 1967. McGraw Hill, New York).

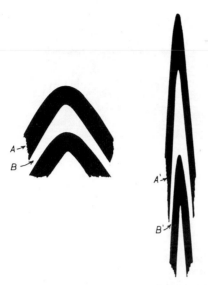

Fig.9.24. Formation of a nearly similar fold A' by the superposition of a uniform compression upon a flexural fold in a competent bed A (Ramsay, 1967. McGraw Hill, New York).

    Lastly, if the deformation takes place in a regime in which
there is a notable component of **coaxial homogeneous flattening**,
all the folds, whatever their mechanism of origin, tend toward a
similar fold shape (fig.9.24). We note that this component makes
the cleavage or foliation approach the axial plane orientation.
We shall see nevertheless in § 9.5 that the role of flattening has
been exaggerated in the interpretation of similar folds.

9.4. FOLDING OF A STRATIFIED SEQUENCE

    Folding typically affects stratified sequences in which the
rheological behaviour determines the operating mechanism and the
type of folds that are produced. Sequences may be made up of beds
of the same competence or of competent beds which are constant or
variable in thickness, alternating with incompetent beds. In the
latter case the bedding has no viscosity contrast and it behaves
as a passive marker during folding.
    Parallel folding of a stratified sequence, which is domi-
nated by competent layers leads theoretically, due to variations
in the radius of curvature of the beds, to a geometrical incompa-
tibility (fig.9.10) on the core of a thick tight fold. The deformation can be accommodated locally by a loss of volume by solution (fig.5.1c). The escape of excess material from the core of a
parallel fold can take place through faults. They can grow from
this core and migrate upwards along the trajectory of limb deformation (fig.9.15) . **Disharmonic folds** developed below the core
of a strong layer are folds without precise geometrical relations
to the principal fold (fig.9.11). They affect incompetent beds
which undergo important flow and indicate a disharmony layer or
decollement (fig.9.25).

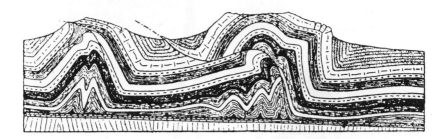

Fig.9.25. Disharmonic folds in the core of a parallel fold and underlying surface of decollement in the Jura chain (Buxtorf, 1916. Verhandl. Naturtorsch. Gesell. Basel, XXVII).

        If the sequence contains a competent bed, much thicker than the others, this bed controls the whole deformation (§9.3.1). Thus the folding of the sub-alpine chain in western Alps is dominated by the existence of two limestone units ; hinge zone deformation accompanied by faulting then predominates (fig.9.14). On the contrary, if the medium is regularly stratified, with competent beds of constant thickness being separated by incompetent layers, as in the case of flysch where there are regular alternations of sandy and shaley beds, flexural slip can predominate, the displacements between the sandstone beds taking place by slip in the interbedded shales. Flow in these interbedded shales from the limbs towards the hinges can be such that the sequence adopts a similar style of folding (fig.9.26a) although the competent beds remain parallel folds. Introducing new material into the hinge zone between such competent beds forms the same type of fold (fig.9.26b). They are called **composite similar folds**. Lastly, if on a given scale the medium can be considered as homogeneous but has a planar anisotropy, deformation results directly in similar folds (fig.9.27).

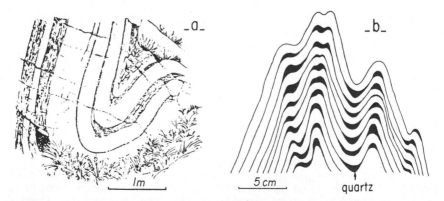

Fig.9.26. Composite similar folds. a) In a regularly stratified sequence where the competent beds deform by parallel folding and the incompetent ones flow towards the hinges. b) In a competent medium infilled along the stratification joints by quartz during folding. (a) Zwart, 1963. Leidse Geol. Med., 28, 321 ; b) Arthaud, 1970. Publ. Univ. Sc. Tech. Languedoc).

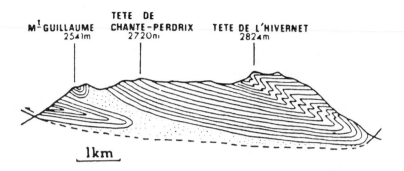

Fig.9.27. Similar folds in flysch. Flysch, a stratified formation
at the outcrop scale, becomes homogeneous on the scale of several
kilometres as considered here (Merle, 1982. Thesis, Rennes).

9.5. RELATIONSHIP OF FOLDS TO THE STRAIN AXES

In Chapter 6 we have seen that in homogeneous deformation
the principal axes of finite strain (X,Y,Z) are simply related to
the structural frame, the (XY) plane coinciding with the cleavage
or foliation and the X or Y axis with the lineation (§6.2.5 and
§6.3.5). In the case of an heterogeneous deformation by folding,
the relation to the fold reference frame generally persists
except in the case of competent beds where the axial plane cleavage forms a convergent fan (§6.2.3).

9.5.1. Relationship between cleavage or foliation and the axial plane

The association of a penetrative cleavage or foliation with
folds can only be seen in the case of similar folding (see
§6.2.6). Commonly the cleavage is very obvious and often strictly
coincides with the axial plane of the folds which is thus the
(X,Y) plane of strain. This observation is not general, for
example figure 6.12 shows that in the neighbourhood of a heterogeneity the path of the (X,Y) plane deviates locally from the
axial plane of the fold. Nevertheless, it militates in favour of
mechanisms calling upon an important component of coaxial flattening in the formation of similar folds. As divergent cleavage
with respect to the axial plane caused by heterogeneous simple
shear is rarely seen, one must conclude that coaxial flattening
plays a major role in the formation of similar folds, acting
alone or in association with buckling or simple shear. This
conclusion must be tempered because simple shear can lead to a
near concordance of the axial plane and the (X,Y) plane if strain
is very large. Thus, if simple shear is considerable as in the
case of a shear zone with $\gamma \geqslant 10$, the angle between these two
planes is $\alpha \leqslant 6°$ which is difficult to discern in the field. Furthermore, the superimposition upon the cleavage of discontinuous
slip surfaces (C planes, § 8.3.1) parallel to the simple shear
plane can alter the appearance and hide any obliquity.

9.5.2. **Relationships between lineations and fold axes**

        This question has already been tackled in §6.3.6 where we
have seen that it is convenient to treat separately lineations
which are essentially parallel to the fold axis (lineations due
to intersections and corrugations) and lineations due to stretching which are generally close to or parallel with the X strain
caxis.
        The direction of the fold axis can be found from intersec-
tion and corrugation lineations, but these lineations supply no
information about finite strain or kinematics. Figures 9.28 for
flexural folds and 9.29 for shear folds show that the fold axes
do not necessarily coincide with the principal strain or movement
axes. In the case of similar folds, the beds are theoretically
passive and do not influence the fold structure. On the contrary
in parallel folds where the competent bed plays an active role,
the fold axis often makes a large angle with the X strain axis as
it is revealed by striations due to sliding on slickensides,
stretching of fossils, pebbles, or reduction spots. In the outer
hinge zone, stretching may appear parallel to the fold axis.

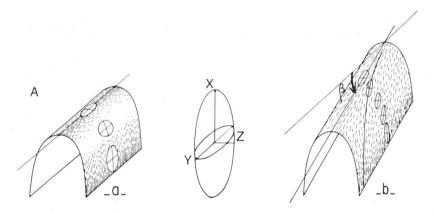

Fig.9.28. Relations between flexural folds and the principal
deformation axes. a) Axis and axial plane parallel to the principal strain axes. b) Oblique axis making an angle β with the Y
axis of the strain ellipsoid (Stainforth, 1978. Tectonophysics,
48, 107).

        Stretching lineations record a strain that can be indepen-
dent of folding. In superficial folds which have been moderately
folded (shortening of a few tens of per cent), one can see that
the stretching is commonly at large angles to the fold axes. On
the contrary, in ductile zones of large strain, stretching lineations are at low angles or parallel to the fold axes (fig.6.17).
This arrangement can be seen in sheath folding (fig.6.17b and
9.30) in which the axes are near to the stretching lineation
except in the zone of conical closure. Such folds can be reproduced experimentally in large scale simple shear. Starting with a
local structural heterogeneity, an instability is generated in
the flow and it manifests itself as an overturned drag fold whose
axis is initially at a high angle with the direction of flow.

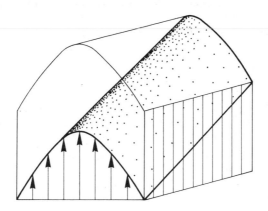

Fig.9.29. Deformation by simple heterogeneous shear of a bed
lying obliquely to the direction of shearing (shear plane :
vertical hachures ; shear direction : vertical arrows). The fold
axis has no relation to the shear direction.

Following rotations due to the large strain (see fig.6.9) the
fold axis is progressively deformed and the sheath structure is
produced (fig.9.30).
    Folds in which the axis is parallel to the X direction, as
in diapirs such as salt domes, may be formed by another process.
During the rise of a diapir, the flow lines tend to converge
(constricted flow or extrusion) in the way that tooth paste comes
out of a tube. All beds initially orientated at a small angle to
the principal flow direction are tightly deformed with their fold
axes parallel to the direction of flow (fig.9.31). In a more

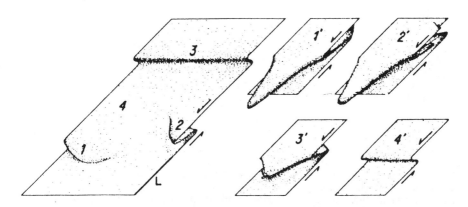

Fig.9.30. Model of the formation of sheath folds in a regime of
simple shear. To the left, different stages of development of
individual folds at a given moment in progressive deformation. To
the right, deformation of these folds (1→1',...) for a more
advanced stage of deformation (Quinquis and Cobbold, 1978. Soc.
géol. France, RAST, 327).

general way, all situations creating a constricted flow can
generate folds in which the axis is parallel to the direction of
flow (fig.9.31b).

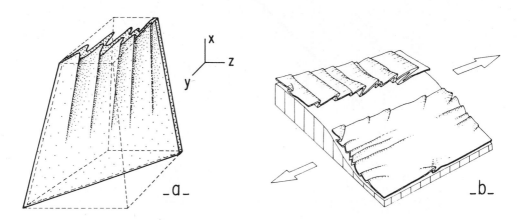

Fig.9.31. Models of the formation of fold axes directly parallel
to the X principal strain axis during constrictive flow (X is
near to the direction of flow and (X,Y), to the flow plane). a)
Mantle diapir. b) Nappes overlying a basement displaced by shearing with a component of gravitational flow causing the constriction. (a) After Nicolas and Boudier, 1975. Tectonophysics, 25,
233 ; b) Van der Driessche, 1983. Thesis, Montpellier).

9.6.SUPERPOSED FOLDS

Metamorphic terrains generally show the imprint of several
episodes of folding. The first episodes are expressed by one or
several folding events contemporaneous with the episodes of
prograde metamorphism (**syn-metamorphic** or syn-schistose folds).
These folds, often similar, isoclinal or intrafolial, reflect the
great ductility of the medium. On the contrary, the last episodes, which occur after the metamorphism, produce more open
folds which are parallel or angular, behaving in a more rigid
fashion. This rheological evolution results from internal changes
in the rocks, such as expulsion of fluid during prograde metamorphism but also from external changes in stress, temperature or
pressure which are related to the decrease in orogenic activity
or in overlying load pressure due to erosion. Subsequent large
scale penetrative deformations occurring in retrograde metamorphic conditions tend to be concentrated in the narrow domains
(shear zones and faults).
Structural analysis of superposed deformations aims at defining the geometrical character of each episode and setting out
their chronology relative to the metamorphic episodes. One attempts to distinguish superposed deformations of the same orogenic event, corresponding to a continuum in tectonic activity,
from deformations that are polyphase, and resulting from independent orogenic events.

9.6.1. Geometrical analysis

Folding is studied by considering the deformation of a surface. We have seen that a major folding is usually accompanied by the formation of a cleavage or foliation nearly parallel to the axial plane of the folds. During a new episode of folding, the initial bedding is refolded leading through interference between the two types of folds to structures that are sometimes so complex that direct analysis is virtually impossible (fig.9.32 and 9.35). On the other hand the foliation only undergoes the second folding. The analysis is much simplified if, after identifying a foliation as the axial plane of the first folding, one uses this foliation as a marker of the second folding (fig.9.33).

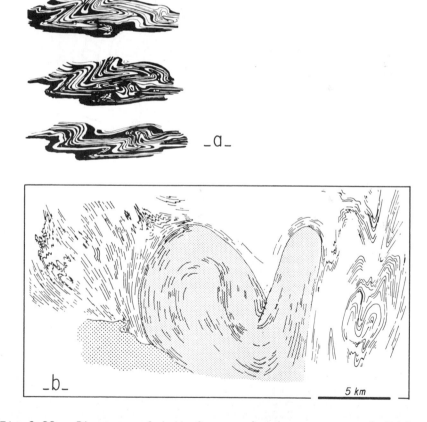

Fig.9.32. Diagrams of interference between superposed folds. a) Experimental folding with conjugate axes with relation to the stretching lineation, seen in the plane perpendicular to it. b) Map of superposed folds in sedimentary and eruptive (grey) formations. (a) Berthé and Brun, 1980, Jour. Str. Geol., 2, 127 ; b) Coward, 1981. Tectonophysics, 76, 59).

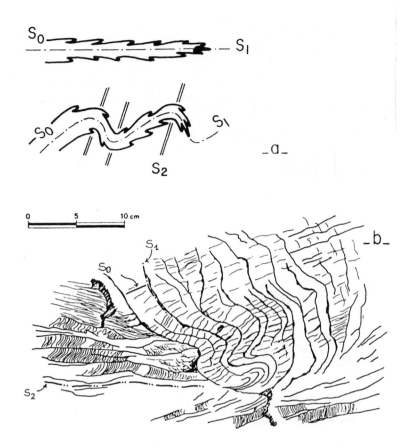

Fig.9.33. Method of analysing superposed folds by using the deformation of the axial plane cleavage or foliation. a) Theoretical diagram : bed So is affected by isoclinal folding which causes an axial plane foliation S1 ; a new folding with axial plane S2 deforms S1 into inclined folds which can be analysed directly, but leads to complex interference patterns if one takes into account the deformation of So. b) Natural example, the folds affecting So are no longer visible (b) G.Gosso, pers. comm.).

This method applies as long as the angle between the axes of the two successive foldings remains small.

Analysis of lineations proves to be more useful when the axes of two successive foldings make large angles. An older lineation reworked obliquely by a more recent flexure fold tends to rotate around the axis of this new fold in a theoretically helicoidal way (fig.9.34) and plots as a small circle distribution on a stereographic projection. On the contrary a more recent lineation imposed obliquely upon an older fold remains in the same plane.

The nature of the folded foliations and associated lineations are useful indicators in the analysis of superposed folds.

Thus a fold affecting a sedimentary bedding in a metamorphic terrain is chronologically earlier than one deforming the foliation produced by this metamorphism. Similarly, a mineral lineation generated during a metamorphic episode is itself older than a corrugation lineation affecting the metamorphic foliation.

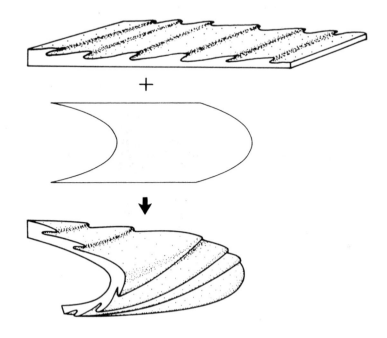

Fig.9.34.   Helicoidal   rolling of an old lineation by a   flexural fold.

9.6.2. Continuum in deformation and polyphase deformations

Some superposed foldings can be attributed to the continuation of the same deformation resulting in the refolding of earlier folds. When two successive foldings share the same style and same orientation, this situation must be suspected (fig.9.35). This condition nevertheless is neither necessary nor sufficient. Folds having different axes can form simultaneously during shearing as in figure 9.32, although in this case the axial plane of the deformation is preserved and the axes are conjugate relative to the direction of stretching. Conversely, deformation can cause a strong **plano-linear anisotropy** which guides later deformation. Folding of a corrugated iron sheet illustrates this situation, as the sheet responds to attempts to bend it in most directions, by bending parallel to the direction of corrugation.

If two superposed foldings of one structural parentage (style, orientation) take place without changes in the mineralogical equilibrium, namely under the same conditions of temperature and pressure, they are thought to correspond to a **deformation continuum**. This is the case in figure 9.35 where during a

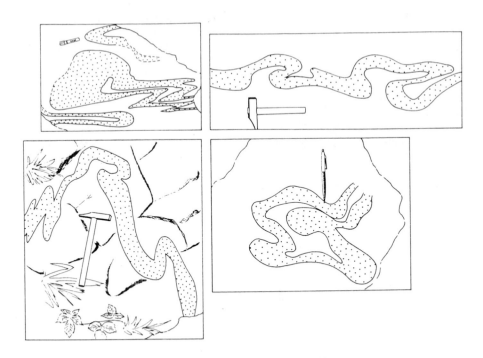

Fig.9.35. Isoclinal similar folds refolded into more open folds with the same axis and axial plane during the diapiric intrusion of peridotites (Nicolas and Boudier, 1975. Tectonophysics, 25, 253).

diapiric intrusion of peridotites in a constrictive regime (§9.5.2), superposed folds were formed which have the same axial plane and axis and also correspond to the same metamorphic facies (plagioclase lherzolite). In every other case, the answer cannot come from structural studies alone.

One can use differences in ductile behaviour between rocks profitably in analysis of a region with superposed deformations. Some particularly competent formations, for example orthogneisses, amphibolites or eclogites, only respond in a ductile manner during the first episodes of syn-metamorphic deformation and do not react to later events apart from fracturing or gentle folding. They may be used for the analysis of the first episodes of deformation which are generally the most relevant in kinematic and geodynamic studies. On the contrary, neighbouring incompetent formations such as micaschists or phyllites continue to behave in a very ductile manner during the later stages, recording them well, but having as a result the record of previous episodes erased. It is necessary therefore, when making correlations of deformation episodes between different formations, to avoid to do so only on the basis of structural analogies (style or orientation of folds). It is not very plausible that an amphibolite adopts the same fold style as a micaschist in the course of a particular episode. For preference one must look to mineral equilibria accompanying the deformation or preceding it, as a guide.

FOR FURTHER READING

Ramsay, J.G., 1967. Folding and fracturing of rocks. McGraw Hill, New York, 562 p.

Vialon, P., Ruhland, M. et Grolier, J., 1976. Eléments de tectonique analytique. Masson, Paris, 118 p.

Whitten, E.H.T., 1966. Structural geology of folded rocks. Rand McNally Chicago, 678 p.

Appendix I

Tensorial Analysis of Strain and Stress

I.1. STRAIN

I.1.1. Tensorial analysis

Let us consider a volume element of an undeformed solid in a system of rectilinear orthogonal axes (fig.I.1a). The distance ds between two nearly points $P(x_i)$ and $P'(x_i+dx_i)$ is given by :

$$ds^2 = (dx_1)^2 + d(x_2)^2 + (dx_3)^2$$

or :

$$ds^2 = \Sigma_i (dx_i)^2$$

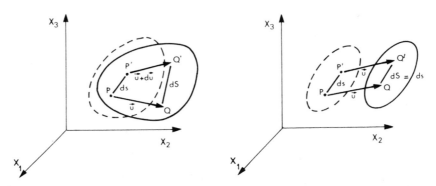

Fig.I.1. a) Deformation : two neighbouring points P,P' lying a distance ds apart are displaced by u and u + du, moving respectively to Q and Q', distant by dS ≠ ds ; b) Rigid body translation : the distance (Q,Q') = (P,P') (Nicolas and Poirier, 1976. Wiley-Interscience, London).

If the points corresponding to P and P' in the deformed solid become Q and Q' respectively, their coordinates in the same system of axes also becomes :

$$X_i = x_i + u_i$$
$$X_i + dX_i = x_i + dx_i + u_i + du_i$$

The distance dS between Q and Q' is :

$$dS^2 = \Sigma_i (dX_i)^2$$

We shall assume that in the following the terms $\partial u_i/\partial x_j$ are infinitely small, that is to say that the vector $PQ = u$ varies slowly with x_i. This vector u is called the **displacement vector** at the point P and its components depend generally on the coordinates of the point P. Together these vectors apply to a domain of finite dimensions defining a **field of displacement.**

Let us imagine firstly the particular case where u is independent of the coordinates of the point P. All the points are then displaced by the same amount ($du_i=0$) and the distance between two neighbouring points does not change. The deformation is then simply reduced to a rigid-body translation (fig.I.1b).

In the general case u depends upon the coordinates of the point P. The vector PP' linking two neighbouring points in the undeformed state becomes QQ' after deformation. One can see on fig.1.1a that its length dS and its orientation to the axes have changed. Thus following a deformation, the local changes in position, orientation, dimension and shape of a volume element depend upon the variation of the displacement vector u as a function of the coordinates of the material points.

Before generalising to the three dimensions, let us look for the physical significance of the components of du in the case of a two dimensional deformation (fig.I.2). Let us consider the

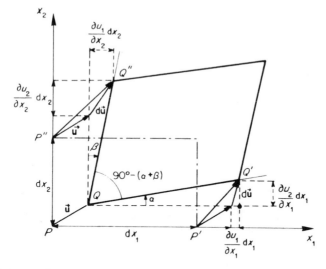

Fig.I.2.   Deformation of an infinitesimal rectangle P,P',P" (broken line) to a parallelogram Q,Q',Q", (solid line) (Nicolas and Poirier, 1976. Wiley-Interscience, London).

deformation of a rectangle of infinitesimal size, defined by three of the corners P,P',P", whose coordinates in a system of axes parallel to the edges of the rectangular are :

P (0,0)
P' (dx1,0)
P" (0,dx2)

One can see on figure I.2 that the segment PP' whose length was initially dx1, becomes QQ'. To a first approximation the length QQ' is :

$$dx_1(1 + \partial u_1/\partial x_1)$$

As a result the quantity :

$$\partial u_1/\partial x_1 \simeq (QQ' - PP')/PP'$$

is the **stretching** or **unit extension** of a segment parallel to the x1 axis. Likewise :

$$\partial u_2/\partial x_2 \simeq (QQ'' - PP'')/PP''$$

is the unit extension of a segment parallel to the x2 axis.
The angle α of QQ' with axis x1 is :

$$\alpha \simeq \tan \alpha \simeq \partial u_2/\partial x_1$$

PP' is parallel to the x1 axis in the undeformed solid. The quantity $\alpha \simeq \partial u_2/\partial x_1$ consequently measures the rotation of a segment initially parallel to this axis during deformation. Similarly, $\beta \simeq \partial u_1/\partial x_2$ measures the rotation of a segment that was initially parallel to the x2 axis.
These considerations, as well as observation of figure 1.2, allow the calculation of the coordinates of points Q, Q', Q'' respectively resulting from the deformation of P, P', P'' :

$$Q \quad \left| \begin{array}{l} u_1 \\ u_2 \end{array} \right.$$

$$Q' \quad \left| \begin{array}{l} dx_1 + u_1 + \dfrac{\partial u_1}{\partial x_1} dx_1 \\[2ex] u_2 + \dfrac{\partial u_2}{\partial x_1} dx_1 \end{array} \right.$$

$$Q'' \quad \left| \begin{array}{l} u_1 + \dfrac{\partial u_1}{\partial x_2} dx_2 \\[2ex] dx_2 + u_2 + \dfrac{\partial u_2}{\partial x_2} dx_2 \end{array} \right.$$

The initial right angle (PP', PP'') is changed by the deformation into $(QQ', QQ'') = 90° - (\alpha + \beta) = 90° - \psi$, with :

$$\frac{\partial u_2}{\partial x_1} + \frac{\partial u_1}{\partial x_2} = \psi$$

The deformation of a rectangle into a parallelogram can be split up into four elementary deformations (fig.I.3).
    1. The edges parallel to the coordinate axes are respectively lengthened by :

$$\frac{\Delta L_1}{L_1} = \frac{\partial u_1}{\partial x_1} \quad \text{and} \quad \frac{\Delta L_2}{L_2} = \frac{\partial u_2}{\partial x_2}$$

The   rectangular shape is retained,   the relative change   in
area being to a first approximation :

$$\frac{\Delta A}{A} \simeq \frac{\partial u_1}{\partial x_1} + \frac{\partial u_2}{\partial x_2}$$

2.   The rectangle is now sheared parallel to each axis by an
angle :

$$\frac{\alpha+\beta}{2} = \frac{1}{2} (\frac{\partial u_1}{\partial x_2} + \frac{\partial u_2}{\partial x_1})$$

This   shearing   changes the rectangle into a parallelogram   wit
the   correct angle Q :   90°− (α+β).   On the other hand the angles
with the coordinate axes are not correct (fig.I.3.3).

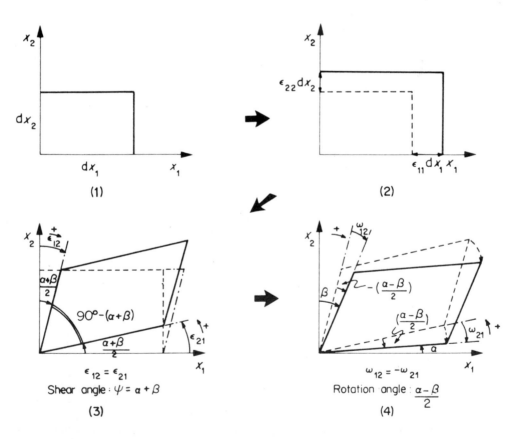

Fig.I.3.   Breakdown in four successive stages of an infinitesimal
deformation   :   1)   Initial state ;   2) Stretching of   the   edges
without change of shape ; 3) Shearing through an angle ψ = α+β ;
4)   Rigid   rotation   through an angle 1/2   (α− β)   (Nicolas   and
Poirier, 1976. Wiley−Interscience, London).

3. The correct orientation of the parallelogram with its axes is found by a rigid rotation of an angle $\omega = (\alpha-\beta)/2$ about the perpendicular axis on Q to the plane of the figure.

These results can be generalised to three dimensions in the general case where the vectors PP' and PP" are not parallel to the coordinate axes. The expression of the vector **du** becomes :

$$du_1 = \frac{\partial u_1}{\partial x_1} dx_1 + \frac{\partial u_1}{\partial x_2} dx_2 + \frac{\partial u_1}{\partial x_3} dx_3$$

$$du_2 = \frac{\partial u_2}{\partial x_1} dx_1 + \frac{\partial u_2}{\partial x_2} dx_2 + \frac{\partial u_2}{\partial x_3} dx_3$$

$$du_3 = \frac{\partial u_3}{\partial x_1} dx_1 + \frac{\partial u_3}{\partial x_2} dx_2 + \frac{\partial u_3}{\partial x_3} dx_3$$

or, in a shortened notation :

$$du_i = \sum_j \frac{\partial u_i}{\partial x_j} dx_j \qquad \begin{array}{l} i = 1,2,3 \\ j = 1,2,3 \end{array}$$

The 9 partial derivatives can be written in the form of a square matrix. These are called the components of the local **strain tensor s.l.**. We have preferred nevertheless, (§2.1) to reserve the term deformation only for changes of shape. The deformation s.l. thus includes deformation in this restricted sense or strain, also called **distortion**, and translation and rotation undergone by the object.

$$\begin{vmatrix} \dfrac{\partial u_1}{\partial x_1} & \dfrac{\partial u_1}{\partial x_2} & \dfrac{\partial u_1}{\partial x_3} \\[2mm] \dfrac{\partial u_2}{\partial x_1} & \dfrac{\partial u_2}{\partial x_2} & \dfrac{\partial u_2}{\partial x_3} \\[2mm] \dfrac{\partial u_3}{\partial x_1} & \dfrac{\partial u_3}{\partial x_2} & \dfrac{\partial u_3}{\partial x_3} \end{vmatrix}$$

The strain tensor s.l. can split up into symmetrical and skew-symmetrical parts. The symmetrical part represents strain (s.s.) and the skew-symmetrical part the rigid rotation.

$$\begin{vmatrix} \dfrac{\partial u_1}{\partial x_1} & \dfrac{\partial u_1}{\partial x_2} & \dfrac{\partial u_1}{\partial x_3} \\[2mm] \dfrac{\partial u_2}{\partial x_1} & \dfrac{\partial u_2}{\partial x_2} & \dfrac{\partial u_2}{\partial x_3} \\[2mm] \dfrac{\partial u_3}{\partial x_1} & \dfrac{\partial u_3}{\partial x_2} & \dfrac{\partial u_3}{\partial x_3} \end{vmatrix} \equiv \begin{vmatrix} \dfrac{\partial u_1}{\partial x_1} & \dfrac{1}{2}(\dfrac{\partial u_1}{\partial x_2} + \dfrac{\partial u_2}{\partial x_1}) & \dfrac{1}{2}(\dfrac{\partial u_1}{\partial x_3} + \dfrac{\partial u_3}{\partial x_1}) \\[2mm] \dfrac{1}{2}(\dfrac{\partial u_2}{\partial x_1} + \dfrac{\partial u_1}{\partial x_2}) & \dfrac{\partial u_2}{\partial x_2} & \dfrac{1}{2}(\dfrac{\partial u_2}{\partial x_3} + \dfrac{\partial u_3}{\partial x_2}) \\[2mm] \dfrac{1}{2}(\dfrac{\partial u_3}{\partial x_1} + \dfrac{\partial u_1}{\partial x_3}) & \dfrac{1}{2}(\dfrac{\partial u_3}{\partial x_2} + \dfrac{\partial u_2}{\partial x_3}) & \dfrac{\partial u_3}{\partial x_3} \end{vmatrix} +$$

$$+ \begin{vmatrix} 0 & \frac{1}{2}(\frac{\partial u_1}{\partial x_2} - \frac{\partial u_2}{\partial x_1}) & \frac{1}{2}(\frac{\partial u_1}{\partial x_3} - \frac{\partial u_3}{\partial x_1}) \\ \frac{1}{2}(\frac{\partial u_2}{\partial x_1} - \frac{\partial u_1}{\partial x_2}) & 0 & \frac{1}{2}(\frac{\partial u_2}{\partial x_3} - \frac{\partial u_3}{\partial x_2}) \\ \frac{1}{2}(\frac{\partial u_3}{\partial x_1} - \frac{\partial u_1}{\partial x_3}) & \frac{1}{2}(\frac{\partial u_3}{\partial x_2} - \frac{\partial u_2}{\partial x_3}) & 0 \end{vmatrix}$$

One can write the components of the strain tensor s.s. in the following condensed form :

$$\frac{\partial u_i}{\partial x_j} = \frac{1}{2}(\frac{\partial u_i}{\partial x_j} + \frac{\partial u_j}{\partial x_i}) + \frac{1}{2}(\frac{\partial u_i}{\partial x_j} - \frac{\partial u_j}{\partial x_i})$$

or again :

$$\frac{\partial u_i}{\partial x_j} = \varepsilon_{ij} + \omega_{ij}$$

The quantities :

$$\varepsilon_{ij} = \frac{1}{2}(\frac{\partial u_i}{\partial x_j} + \frac{\partial u_j}{\partial x_i})$$

can be written in the form of a square matrix :

$$\begin{vmatrix} \varepsilon_{11} & \varepsilon_{12} & \varepsilon_{13} \\ \varepsilon_{21} & \varepsilon_{22} & \varepsilon_{23} \\ \varepsilon_{31} & \varepsilon_{32} & \varepsilon_{33} \end{vmatrix}$$

Given that the tensor is symmetrical, $\varepsilon_{ij} = \varepsilon_{ji}$, the number of independent components is reduced to 6. At a given point in the solid, these six components of the deformation include all the information regarding the change of shape and size of an elementary volume centred on this point.

A diagonal component of this matrix ε_{ii} represents the **unit extension** of a segment parallel to the x_i axis. A off-diagonal component ε_{ij} ($i \neq j$) represents half the **shear angle** ψ which changes a rectangle into a parallelogram in the x_ix_j plane. These shear strains are responsible for changes in shape of the initial elementary volume at constant volume. The change in volume is given by :

$$\frac{\Delta V}{V} \simeq \sum_i \varepsilon_{ii} = \varepsilon_{11} + \varepsilon_{22} + \varepsilon_{33}$$

By a judicious choice of the system of axes one can always make the strain tensor diagonal, thus cancelling the non-diagonal components. These particular axes define the **principal directions of strain**.

The quantities :

$$\omega_{ij} = \frac{1}{2}(\frac{\partial u_i}{\partial x_j} - \frac{\partial u_j}{\partial x_i})$$

are the components of the **rotation tensor** expressing the **rigid rotation** or **vorticity** of the elementary volume considered.

I.1.2. Typical strain regimes

Homogeneous strain. If all the partial derivatives of the vector du are constant in the whole volume, that is to say they are independent of the coordinates of the different points, the components of the strain tensor are the same at every point, the strain is uniform or homogeneous.

Coaxial and non-coaxial strain. In a condition of uniform strain, if the tensorial components of rotation are zero at all points, the strain is coaxial ; if not it is non-coaxial.

Plane strain. All the displacement vectors u are parallel to a plane and therefore do not depend on the third coordinate, for example x3, therefore :

$$\begin{vmatrix} u3 = 0 \\ u2u1 \text{ are independent of } x3 \end{vmatrix}$$

Pure and plane shear. This is coaxial strain (fig.I.4a). Then :

$$\frac{\partial u_1}{\partial x_2} = \frac{\partial u_2}{\partial x_1} \quad \text{and} \quad \begin{vmatrix} \varepsilon_{12} = \varepsilon_{21} = \frac{\psi}{2} \\ \omega_{12} = -\omega_{21} = 0 \end{vmatrix}$$

Simple shear. This is a plane deformation (fig.1.4c) in which :

$$\frac{\partial u_1}{\partial x_2} = \psi \quad \frac{\partial u_2}{\partial x_1} = 0 \qquad \begin{vmatrix} \varepsilon_{12} = \varepsilon_{21} = \frac{\psi}{2} \\ \omega_{12} = \frac{\psi}{2} \\ \omega_{21} = -\frac{\psi}{2} \end{vmatrix}$$

Because it is infinitesimal, this deformation can thus be considered as the sum of a pure strain corresponding to a shear of ψ
and a rigid rotation at an angle $\psi/2$ (fig.I.4).

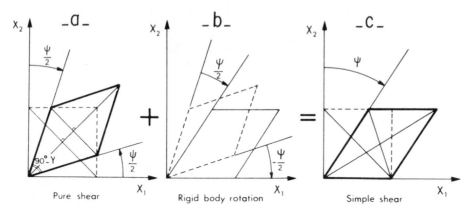

Fig.I.4. Breakdown of simple shear through an angle ψ (c) into
(a) pure shear, plus (b) rigid body rotation at an angle of $\psi/2$
(Nicolas and Poirier, 1976. Wiley-Interscience, London).

Isotropic and deviatoric strain. The strain tensor s.s. can be
divided into two parts : a isotropic part which involves dilatation or compression (positive or negative volume change) and a
deviatoric part responsible for changes in shape. We have seen
that the total volume change can be given by :

$$\frac{\Delta V}{V} \simeq \varepsilon_{11} + \varepsilon_{22} + \varepsilon_{33}$$

This change in volume can be achieved by a strain tensor in
which each diagonal term equals $1/3(\varepsilon 11 + \varepsilon 22 + \varepsilon 33)$; the sum of the
three diagonal terms will equal $\Delta V/V$. The deviatoric part is
given by the difference between the strain tensor and its isotropic part. The deviatoric part represents strain at constant
volume, the sum of the diagonal terms is thus zero.
This can be written :

$$
\begin{vmatrix} \varepsilon_{11} & \varepsilon_{12} & \varepsilon_{13} \\ \varepsilon_{21} & \varepsilon_{22} & \varepsilon_{23} \\ \varepsilon_{31} & \varepsilon_{32} & \varepsilon_{33} \end{vmatrix} \equiv
\begin{vmatrix} \frac{1}{3}(\varepsilon_{11}+\varepsilon_{22}+\varepsilon_{33}) & 0 & 0 \\ 0 & \frac{1}{3}(\varepsilon_{11}+\varepsilon_{22}+\varepsilon_{33}) & 0 \\ 0 & 0 & \frac{1}{3}(\varepsilon_{11}+\varepsilon_{22}+\varepsilon_{33}) \end{vmatrix} +
$$

$$
+
\begin{vmatrix} \varepsilon_{11} - \frac{1}{3}(\varepsilon_{11}+\varepsilon_{22}+\varepsilon_{33}) & \varepsilon_{12} & \varepsilon_{13} \\ \varepsilon_{21} & \varepsilon_{22} - \frac{1}{3}(\varepsilon_{11}+\varepsilon_{22}+\varepsilon_{33}) & \varepsilon_{23} \\ \varepsilon_{31} & \varepsilon_{32} & \varepsilon_{33} - \frac{1}{3}(\varepsilon_{11}+\varepsilon_{22}+\varepsilon_{33}) \end{vmatrix}
$$

I.2. STRESS

I.2.1. Tensorial analysis

Let us consider a solid continuum in equilibrium under applied forces. These can be divided into two categories ; a) body forces per unit volume such as gravity, b) surface forces applied to the external surface of the solid, F per unit area. One can understand the idea of stress by the following imaginary experiment. Within the solid we define a volume of matter V bounded by a surface S and isolate it ; in order for the volume to remain in equilibrium it is necessary to apply forces upon the surface S. These forces represents the action that the surrounding solid exerts upon the volume V.

We now consider an elementary area dA of S centered on point O, with the normal to S being oriented positively outward. The action exerted by the surrounding solid upon the volume V at O can be reduced to a force TdA which is proportional to dA. The **stress vector T** relative to the elementary area centered on O is defined as the force per unit area of surface exerted by the exterior solid (situated on the side of the positive normal) upon the part of the solid situated inside this surface. Let us now consider an elementary tetrahedron OABC of volume dV defined as follows : O is the origin of the coordinates and the three orthogonal edges OA, OB, OC coincide with the coordinate axes (fig.1.5). If dA is the area of the face ABC, the surfaces of the

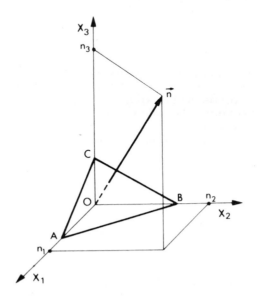

Fig.I.5. Elementary tetrahedron OABC. n1, n2, n3 : projections along the axes of vector n normal to the face ABC (Nicolas and Poirier, 1976. Wiley-Interscience, London).

faces lying perpendicular to the axes Ox1, Ox2,OX3 are respectively :

> dA1 = n1dA
> dA2 = n2dA
> dA3 = n3dA

where the ni's (i = 1,2,3) are the components of n, the unit
vector normal to the face ABC.
    We designate the stress vector relative to the face ABC by
T, and T',T",T"' as the stress vectors normal to Ox1, Ox2,Ox3.
The tetrahedron separated from the solid will be in equilibrium
under the surface forces :

+TdA, -TdA1,-T"dA2, -T'"dA3

and the volume force :

+ φ dV

    The equilibrium condition is that the sum of the forces be
equal to zero :

TdA-T'n1dA-T"n2dA-T'"n3dA + φ dV = 0

or:

T - T'n1 - T"n2 - T'"n3 + φ dV/dA = 0

    If the dimensions of the tetrahedron tend to zero, dV/dA
tends to zero and the last term on the left can be neglected. The
components of the stress vectors become (fig.1.6) :

$$T \begin{vmatrix} T_1 \\ T_2 \\ T_3 \end{vmatrix} \qquad T' \begin{vmatrix} \sigma_{11} \\ \sigma_{21} \\ \sigma_{31} \end{vmatrix} \qquad T'' \begin{vmatrix} \sigma_{12} \\ \sigma_{22} \\ \sigma_{32} \end{vmatrix} \qquad T''' \begin{vmatrix} \sigma_{13} \\ \sigma_{23} \\ \sigma_{33} \end{vmatrix}$$

    The equilibrium condition projected upon the coordinates is
then written :

$$\begin{vmatrix} T_1 = \sigma_{11}n_1 + \sigma_{12}n_2 + \sigma_{13}n_3 \\ T_2 = \sigma_{21}n_1 + \sigma_{22}n_2 + \sigma_{23}n_3 \\ T_3 = \sigma_{31}n_1 + \sigma_{32}n_2 + \sigma_{33}n_3 \end{vmatrix}$$

    These three equations can be written synoptically as :

$T_i = \Sigma_j \sigma_{ij} n_j$

    These equations show that the **stress vector** at the point O
(fig.I.6) relative to a surface element defined by n can be
expressed as a function of 9 quantities σ_{ij} called the **stresses**
at point O. These quantities entirely define the state of stress
at a given point in the solid.

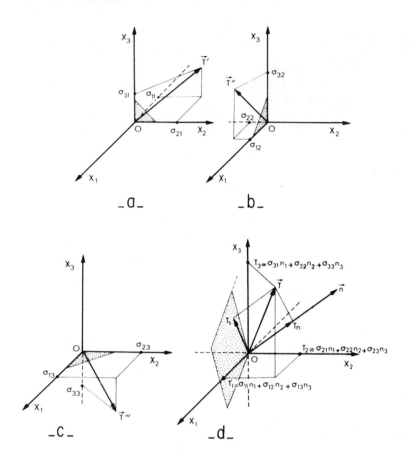

Fig.I.6. Definition of the stresses : σij is the projection along
xi axis of the stress vector relative to the face (stippled)   of
the elementary tetrahedron normal to the xj axis.   a) **T'** : stress
vector to the face normal to x1.   b) **T"** :   stress vector relative
to the face normal to x2.   c) **T"'** : stress vector relative to the
face normal to x3.   d) **T** :   stress vector relative to the   face
normal to n (Nicolas and Poirier, 1976. Wiley-Interscience,
London).

        We   note that the stress σij is the component of the stress
vector at O along the xi axis, in relation to the surface element
perpendicular   to the xj axis.   It can be shown that   in   general
$\sigma ij = \sigma ji$;   the   9 components of stress are reduced then   to   6
independent   components   that   one can write as a   square   matrix
representing the stress tensor :

$$\begin{vmatrix} \sigma_{11} & \sigma_{12} & \sigma_{13} \\ \sigma_{21} & \sigma_{22} & \sigma_{23} \\ \sigma_{31} & \sigma_{32} & \sigma_{33} \end{vmatrix}$$

The value of each of these components depends on the choice of axes. On the contrary the quantity $P = \sigma_{11} + \sigma_{22} + \sigma_{33}$ is independent of the choice of these axes. $P/3$ represents the **isotropic pressure** at the point 0, for the stress tensor can be divided into an isotropic part and a deviatoric part, as in the case of the strain tensor :

$$
\begin{vmatrix} \sigma_{11} & \sigma_{12} & \sigma_{13} \\ \sigma_{21} & \sigma_{22} & \sigma_{23} \\ \sigma_{31} & \sigma_{32} & \sigma_{33} \end{vmatrix} = \begin{vmatrix} \dfrac{P}{3} & 0 & 0 \\ 0 & \dfrac{P}{3} & 0 \\ 0 & 0 & \dfrac{P}{3} \end{vmatrix} + \begin{vmatrix} \sigma_{11} - \dfrac{P}{3} & \sigma_{12} & \sigma_{13} \\ \sigma_{21} & \sigma_{22} - \dfrac{P}{3} & \sigma_{23} \\ \sigma_{31} & \sigma_{32} & \sigma_{33} - \dfrac{P}{3} \end{vmatrix}
$$

The stress vector which has been broken down along the coordinate axes as a function of the σ_{ij} components, can also be broken down in relation to a surface element and its normal. The projection on the surface plane is called the **shear stress** (Ts) at point 0 on that plane and the projection along the normal **n** to this surface element is the **normal stress** (Tn) (fig.I.6). Ts and Tn can also be expressed as functions of the σ_{ij} components. One can see also that the diagonal components correspond to the normal stresses on planes which are themselves perpendicular to the coordinate axes, whereas the off-diagonal components corres- pond to shear stresses on these planes.

I.2.2. **Typical stress regimes**

Uniaxial traction and compression. Along the Ox3 axis, they are defined by (fig.I.7) :

$$
\begin{vmatrix} 0 & 0 & 0 \\ 0 & 0 & 0 \\ 0 & 0 & \sigma_{33} \end{vmatrix}
$$

The stress vectors on the planes at right angles to the different axes are :

$$
T' \begin{vmatrix} 0 \\ 0 \\ 0 \end{vmatrix} \qquad T'' \begin{vmatrix} 0 \\ 0 \\ 0 \end{vmatrix} \qquad T''' \begin{vmatrix} 0 \\ 0 \\ \sigma_{33} \end{vmatrix}
$$

The regime corresponds to an uniaxial traction if σ_{33} is positive and to an uniaxial compression if this quantity is negative.

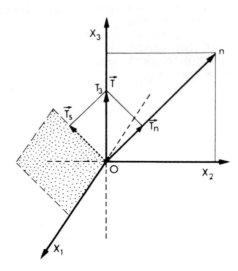

Fig. I.7. Uniaxial traction along the Ox3 axis. T : stress vector on a plane at 45° from Ox3 (stippled) ; Tn : normal stress ; Ts : shear stress resolved on this plane (Nicolas and Poirier, 1976, Wiley-Interscience, London).

Only the plane x1Ox2 is under normal stress. Planes x1Ox3, x2Ox3 are not stressed.

Let us consider any plane whose normal makes an angle of 45 with Ox3 (n3 = cos 45° = $\sqrt{2}/2$). The stress vector on such a plane has as its components (fig.I.7) :

$$T \begin{vmatrix} T_1 = 0 \\ \\ T_2 = 0 \\ \\ T_3 = n_3 \sigma_{33} = \sqrt{\frac{2}{2}} \sigma_{33} \end{vmatrix}$$

We can easily calculate the normal stress relative to this plane : Tn = T3cos45° = $\sigma_{33}/2$. The shear stress created or resolved on this plane Ts is equal to Tn. It is also apparent that all planes making an angle of 45° with the axis of tension or compression have the **maximum resolved shear stress.**

Isotropic tension or compression. This regime is one of compression if the stress is negative and of tension if it is positive. These regimes are defined as :

$$\begin{vmatrix} \sigma_{11} & 0 & 0 \\ \\ 0 & \sigma_{22} & 0 \\ \\ 0 & 0 & \sigma_{33} \end{vmatrix}$$

The stress vector **T** on a plane defined by its normals (n1,n2,n3) has the following components (fig.I.6) :

$$
\mathbf{T} \quad
\begin{vmatrix}
\text{T1} = \sigma \, \text{n1} \\
\text{T2} = \sigma \, \text{n2} \\
\text{T3} = \sigma \, \text{n3}
\end{vmatrix}
$$

The stress vector is always normal to this plane. There is no shear stress. Figure I.8 illustrates a few axisymmetric stress regimes.

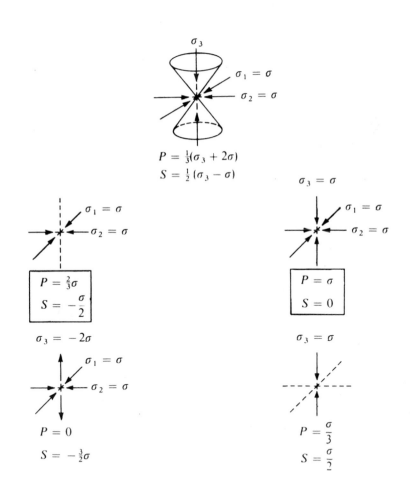

Fig.I.8. Typical axisymmetric regimes of stress. P : isotropic pressure ; S : maximum shear stress on planes at 45° from $\sigma 3$ (tangent to cone) (Nicolas and Poirier, 1976. Wiley-Interscience, London).

I.3. ELASTIC DEFORMATION

Elastic deformation has been defined in § 2.2.2. as a reversible deformation expressed by a linear relationship between stress and strain in the case of uniaxial compression or tension. In the most general case, corresponding to the deformation of an anisotropic solid under the effects of three different principal stresses, this relationship becomes :

$$\sigma_{ij} = C_{ijkl} \, \varepsilon_{kl}$$

The σ_{ij} are the stress components defined above and the ε_{kl}, the components of the infinitesimal strain tensor. The coefficients C_{ijkl} are material constants that only depend upon the physical properties of the solid. These are called the **elastic constants**. To illustrate this, we write the relation between the applied stress σ_{11} and the 9 components of the strain tensor:

$$\sigma_{11} = C_{1111}\varepsilon_{11} + C_{1112}\varepsilon_{12} + C_{1113}\varepsilon_{13} + C_{1121}\varepsilon_{21} + C_{1122}\varepsilon_{22} +$$

$$+ C_{1123}\varepsilon_{23} + C_{1131}\varepsilon_{31} + C_{1132}\varepsilon_{32} + C_{1133}\varepsilon_{33}$$

In the most general case, we write the 9 equations of this type relating the 9 components to each tensor. There are therefore 81 elastic constants.
Taking account of the symmetry of the strain and stress tensors :

$$C_{ijkl} = C_{klij}$$

$$C_{ijkl} = C_{jilk}$$

Thus in the most general case the 81 elastic constants can be reduced to 21. For crystals with the lowest symmetry (triclinic), none of these 21 constants is zero. The number of independent constants decreases as the crystallographic symmetry increases. Thus the cubic system has only three non-zero independent elastic constants : C1111, C1122 and C2323 which using the correspondance rule can be written : C11, C12 and C44. This is not an isotropically elastic solid as Young Modulus depends upon the direction considered. In this regard, crystalline aggregates without preferential orientations constitutes an isotropic solid. There is then a relationship between the three elastic constants and the number of independent constants drops to two. One can write :

$$\left|\begin{array}{l} C_{12} = \lambda \\ \\ C_{44} = \mu \end{array}\right. \qquad C_{11} = \lambda + 2\mu$$

λ and u are called **Lamé coefficients**. In introducing these coefficients in the definition relationship one obtains :

$$\left| \begin{array}{l} \sigma_{11} = \lambda(\varepsilon_{11} + \varepsilon_{22} + \varepsilon_{33}) + 2\mu\varepsilon_{11} = \lambda \dfrac{\Delta V}{V} + 2\mu\varepsilon_{11} \\[2mm] \sigma_{12} = 2\mu\varepsilon_{12} \\[2mm] \sigma_{13} = 2\mu\varepsilon_{13} \\[2mm] \sigma_{22} = \lambda(\varepsilon_{11} + \varepsilon_{22} + \varepsilon_{33}) + 2\mu\varepsilon_{22} = \lambda \dfrac{\Delta V}{V} + 2\mu\varepsilon_{22} \\[2mm] \sigma_{23} = 2\mu\varepsilon_{23} \\[2mm] \sigma_{33} = \lambda(\varepsilon_{11} + \varepsilon_{22} + \varepsilon_{33}) + 2\mu\varepsilon_{33} = \lambda \dfrac{\Delta V}{V} + 2\mu\varepsilon_{33} \end{array} \right.$$

The proportionality coefficient μ between the shear stresses and shear strains is called the **shear modulus** (often designated as G). The proportionality coefficient between the isotropic pressure $P = 1/3(\sigma 11 + \sigma 22 + \sigma 33)$ and the resulting dilatation $-\Delta V/V = \varepsilon 11 + \varepsilon 22 + \varepsilon 33$ is called the **incompressibility** or **Bulk Modulus B** (it is the opposite of compressibility). It can be shown that :

$$B = - \frac{P}{\Delta V/V} = \frac{1}{3}(\lambda + \mu)$$

We can see that the μ modulus controls the changes in shape and the B modulus the changes in volume for a given stress during elastic deformation.

Lastly, one can relate Young Modulus (§2.2.2) to Lamé coefficients and introduce a new coefficient, **Poisson ratio** ν.

$$Y = \frac{\mu(3\lambda + 2\mu)}{\lambda + \mu}$$

$$\nu = \frac{\lambda}{2(\lambda + \mu)}$$

In uniaxial traction or compression of an isotropic body, Young Modulus explains the relationship between deviatoric stress and longitudinal change of length and Poisson ratio the relationship of transverse to longitudinal strain. Poisson ratio is equal to 0.5 in an ideal incompressible solid ($B = \infty$) ; it is in actual solids, commonly with a value around 0.25 which corresponds to the condition $\lambda = \mu$ (Poisson relation).

Appendix II

Measurement of Finite Strain
in collaboration with J.P. Brun and P. Choukroune

II.1. INTRODUCTION

The transformation of rocks, from an initial undeformed to a final deformed state, can theoretically be described by means of a complex geometrical operation in which one can decompose the translation, rigid rotation and strain or distortion (appendix I). Only the strain to which we have restricted the term deformation, is identifiable in the rock. It is measurable provided that the rock encloses as markers, reference objects whose initial shapes are individually or statistically known. We shall here limit the analysis to the measurement of homogeneous finite strain with the aid of these markers and to a review of the basic principles and current methods. Strain being characterized by an ellipsoid (§2.1), the only objective pursued so far in the book was the recognition of the **directions of the X, Y, Z axes** (§6.2.5, 6.3.5, 9.5). The present objective includes the absolute or relative measurement of **the magnitude of these axes.**

II.2. DEFINITIONS

Homogeneous finite deformation can be described in three dimensions by the change of a sphere of unit radius into a strain ellipsoid or, in two dimensions, by that of a circle into an ellipse (§2.1.2).

II.2.1. Strain of a linear marker

If Lo is the initial length of a line and L1 is its final length, the following parameters can be defined :
- Elongation $\lambda = (1 + e)$ with $e = (L1-Lo)/Lo$ (§2.2.2)and the logarithmic or natural strain $\varepsilon = \ln \lambda$ which is different from ε L defined in § 2.2.2 ;
- Quadratic elongation $\Lambda = \lambda^2$;
- Shear strain $\gamma = \tan \psi$ where ψ measures the deformation of a right angle that is positive if the rotation is clockwise (fig.II.1).

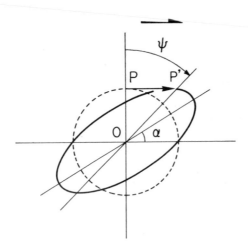

Fig.II.1.   Simple   shear of a line OP to OP' and of a circle into
an ellipse.   The shear strain is γ = PP'/OP and ψ is the   shear
angle.

II.2.2. **Strain in simple shear**

Figure   II.1 describes the principal geometrical   properties
of simple shear (§2.4). The first increment in progressive deformation   changes a circle into an ellipse in which the large   axis
is orientated at 45° to the trace of the shear plane.   The following   increments   approach the two directions   separated   by   the
angle α . There is a relationship between the shear angle ψ and α :

γ = tan ψ = 2 cotan2α

Similarly   in a transformation keeping the surface   constant
($\lambda 1 \lambda 2$ = 1),   one can establish the following relationship between the angle α that can be measured in favorable cases (§II.5)
and   the strain of the corresponding   ellipse whose axes are λ 1
and λ 2 :

γ = $\lambda 1$ - λ 2 = 2 cotan2α

II.2.3. **Strain ellipsoid**

Developments   in this appendix lead to a definition   of   the
strain ellipsoid,   by substituting the values λ 1$\geqslant \lambda$ 2 $\geqslant \lambda$ 3, as defined   above (§II.2.1) for the direct measurements of the axes of
the ellipsoid,   X $\geqslant$ Y $\geqslant$ Z (§2.1.2).   The shape of the strain ellipsoid is completely defined by two parameters,   K   introduced   in
§2.1.3   and   an "intensity" parameter like the ε s   parameter   of
Nadai (1963) :

$$\varepsilon_s = \frac{\sqrt{3}}{2} \gamma_0$$

where : $\gamma_0 = (2/3)\left[(\varepsilon_1-\varepsilon_2)^2 + (\varepsilon_2-\varepsilon_3)^2 + (\varepsilon_3-\varepsilon_1)^2\right]^{1/2}$

with $\varepsilon i = \ln \lambda i, i = 1,2,3$

Five   types   of   ellipsoids   can   now   be   distinguished corresponding to five states of finite strain :

K = 0 Uniaxial flattening ; $\lambda 1 = \lambda 2 > 1$
0<K<1 Triaxial flattening ; $\lambda 1 > \lambda 2 > 1$
K = 1 Plane strain ; $\lambda 2 = 1$
1< K<∞ Triaxial   constriction; $1 > \lambda 2 > \lambda 3$
K = ∞ Uniaxial constriction ; $1 > \lambda 2 = \lambda 3$

Flinn's   diagram   (fig.2.5) where $\lambda 1/\lambda 2$ is expressed as   a function   of $\lambda 2/\lambda 3$ shows the relationships between these   five types   of deformation.   In this diagram the more distant a   point becomes   from its origin the more intense is the strain   (increasing values of the isodeformational curves ε s).

II.2.4. Strain ellipse

The   state of strain on a plane is described by the shape of the   strain   ellipse and the orientation of   its   principal   axes $\lambda G > \lambda P$.   The   shape of the ellipse can be described by the   axial ratio $\lambda G/\lambda P$ (varying between 1 and ∞) or $\lambda P/\lambda G$ (varying   between   1   and 0).   The change in area : $\Delta S = \lambda G \lambda P - 1$,   is   the difference of area ΔS between the ellipse and the initial   circle of   unit radius.   All forms of strain ellipses can be represented on a diagram $\Lambda G \Lambda P$ (fig.II.2).   The initial circle is at $\Lambda G = \Lambda P$ =1.   The line joining this point to the origin corresponds to the

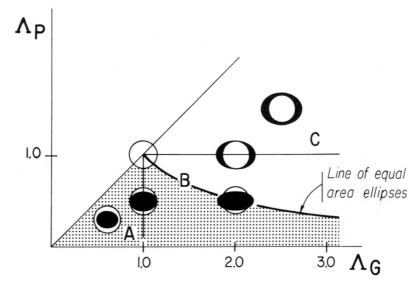

Fig.II.2.   Classification   of   strain   ellipses.   The   field   of ellipses with a reduced surface is shaded.

circles with a radius of less than 1 between the point and the
origin, and to the circles of radius greater than 1, between the
point and infinity. Line $\Lambda G = 1$ corresponds to ellipses resulting from uniaxial shortening and line $\Lambda P = 1$, to ellipses from
uniaxial stretching. The three fields defined by these two lines
correspond to ellipses contained by the initial circle (A),
intersecting the initial circle (B) or containing the initial
circle (C).

The ellipses of field (B) are characterised by two lines
which are not lengthened ($\lambda=1$) (fig.II.3), which separate the
zone of stretched lines from that of shortened lines.

The curve defined by $\Lambda P = 1/\Lambda G$ corresponds to ellipses of
equal area to that of the initial circle ($\Delta S= 0$) ; it lies in the
B field and separates enlarged ellipses from diminished ones.

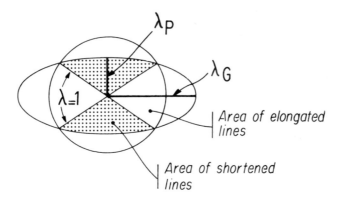

Fig.II.3. Strain ellipse, fields of elongation (white) and
shortening (dotted) separated by invariant lines $\lambda = 1$.

II.2.5. **Mohr circle of strain**

Every line making an angle θ' with λG is characterised by a
reciprocal quadratic elongation :

$$\Lambda'' = \Lambda'G\cos^2\theta' + \Lambda'P\sin^2\theta' \quad (1)$$

and a shear :

$$\gamma' = (\Lambda'P- \Lambda'G)\sin\theta'\cos\theta' \quad\quad (2)$$

where the quadratic reciprocal extensions Λ', $\Lambda'G$ and $\Lambda'P$, are
the inverse of the quadratic extensions such that $\Lambda'' = 1/\Lambda$, or
$\gamma'=\gamma/\Lambda$. The second equation shows that the shear is zero, θ' then
being equal to $0°$ and $90°$, along the principal λG and λP axes.
These two equations are basic to numerous methods of finding the
strain ellipse, which we shall illustrate in a few cases. They
can be put in the form :

$$\Lambda' = 1/2(\Lambda'_G + \Lambda'_P) - 1/2(\Lambda'_P - \Lambda'_G)\cos 2\theta'$$

$$\Lambda' = 1/2(\Lambda'_P - \Lambda'_G)\sin 2\theta'$$

that is to say under the classic form of parametric equations of
a circle :

$$x = c + r' \cos\alpha$$
$$y = r'\sin\alpha$$

This offers the possibility, as for stresses (§2.3.2), of
resolving some of the problems posed by strain analysis by the
graphical Mohr construction. In the present case, the abscissa
axis corresponds to the reciprocal quadratic extensions Λ' and
the ordinate axis to shear strain γ' (fig.11.4). The centre of
the circle is given by half the sum of reciprocal principal
quadratic extensions and its radius by half their difference. The
angular shearing ψ along a line making an angle θ' with $\lambda'G$ is
directly obtained by joining the point P on the circle to the
point of origin of the diagram.

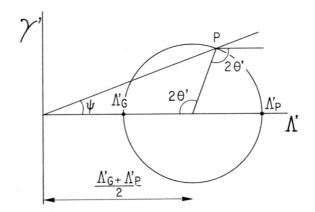

Fig.II.4. Mohr circle of strain.

II.3. DETERMINATION OF THE STRAIN ELLIPSE

All methods alluding to the measurement of the strain ellipsoid or ellipse lead to a result that is only strictly correct
for the objects being considered. The corresponding measurements
apply to the matrix of these objects and in consequence to the
overall medium, if there is no viscosity contrast between the
object and its matrix. If the object is less deformable, as is
frequently the case, the measurement obtained is less than that
of the overall strain.

II.3.1. Measurement of elongation

When one can measure the elongation according to several
orientations upon a plane, it is easy to estimate the principal
elongations and their orientation. This type of measurement is
particularly suited to objects that have been pulled apart by
deformation such as fossils (belemnites, crinoids, etc...) or
minerals (tourmaline, amphibole etc...).

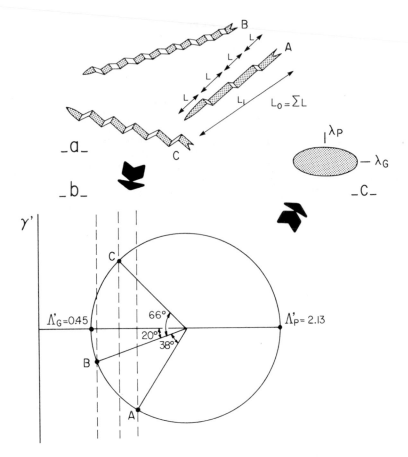

Fig.II.5. Determination of the strain ellipse starting from three
measurements of elongation. a) Boudinaged belemnites. b) Mohr
diagram. c) Corresponding strain ellipse (Modified from Ragan,
1968. Wiley, London).

Let us consider the case of three deformed belemnites
(fig.II.5a). The sum of the lengths of each segment gives Lo, the
total length of the broken belemnite L1. Thus for each belemnite
one can calculate a value for Λ' which is in our example
$\Lambda'A$ =0.86 , $\Lambda'B$ = 0.50 and $\Lambda'C$ = 0.69. The angles between the
belemnites are 19° between A and B and 33° between B and C. The
Mohr construction is shown as follows (fig.II.5b) : the three
directions are drawn on tracing paper at angles $2\theta'$ apart, say 38°
and 66° and one traces several circles of whatever radii, that
have a coincident central point. On a diagram γ' Λ' where the
axis of Λ' is graduated, one traces three vertical lines for the
three values of Λ' that were obtained. Then, by superposition of
the tracing and the γ' Λ' diagram, one looks for the best fit
between the values of $2\theta'$ and Λ' for A,B and C. After a prelimi-
nary adjustment the real value of the radius of the circle is
quickly found. The intersections of the circle with the Λ' axis
give $\Lambda'G$ = 0.45 and $\Lambda'P$= 2.13 and their orientation (in $2\theta'$) in
relation to A, B and C. The axial ratio of the ellipse is
$\lambda G/\lambda P$=2.19.

II.3.2. Measurement of shear strain

When the shear strain along diversely orientated lines upon a plane can be measured, it is possible to estimate the state of strain in this plane. The shear strain is measurable directly if there are objects which have a known right angle. This is the case for many fossils such as brachiopods, trilobites, etc. The method can also be adapted to objects having any angle that is known initially.

Let us consider the example of brachiopods (fig.II.6). The hinge line is chosen as a reference line. The shear angle ψ is the angle which separates the axis of their median sinus from the perpendicular to the hinge. In the case in figure II.6, $\psi A =-13°$, $\psi B =+ 20°$ and $\psi C =+14°$, and the angles θ between B and C and between A and C are 17.5° and 53°. On tracing paper,

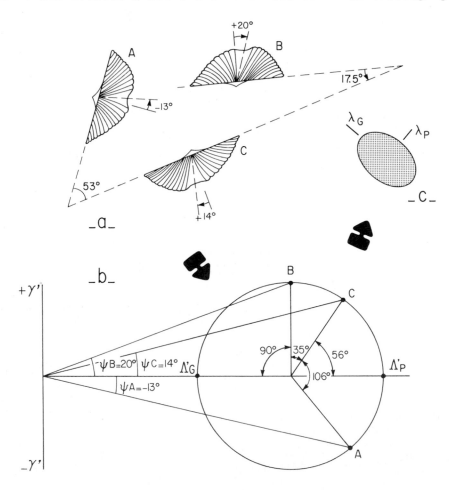

Fig.II.6. Determination of the strain ellipse from three measurements of shear strain. a) Deformed brachiopods. b) Mohr diagram. c) Corresponding strain ellipse (Modified from Ragan, 1968, Wiley, London).

one draws as before the three concurrent directions as double angles (2θ'), as 35° and 106° and a circle of any radius. On a diagram γ'Λ', one draws three straight lines corresponding to the three measured angles. In superimposing the tracing paper on the diagram γ'Λ': the centre of the circle lies on the Λ' axis and by turning the circle one can find the best fit between the straight lines and 2θ' for A, B, and C. As one does not know an absolute value for λ , it is only possible here to obtain the axial ratio of λ G/ λ P = (Λ'P/Λ'G) = 1.48. The orientation of the principal axes λ G and λ P is found as before ; λ P makes an angle of 28° with C in the acute angle between A and C and λ G makes an angle of 45° with B.

II.3.3. Deformed elliptical objects

There are many objects in rocks which have elliptical or semi-elliptical cross-sections. These are for example, pebbles in a conglomerate, reduction spots in slate, spots in contact meta-morphosed rocks, inclusions in granite etc. All ellipses with an initial axial ratio Ri are changed by homogeneous deformation into another ellipse with a final axial ratio Rf. If in the initial state the elliptical objects had their long φi axes in various orientations to any reference line, it is possible, knowing a number of pairs of Rf φf values in the final state , to work out the shape and orientation of the strain ellipse. This method, called the "Rf/φ" method, was devised by Ramsay (1967) and is the one most commonly used to measure finite strain.

Let us consider a group of ellipses numbered from 1 to 10 (fig.II.7a) with constant Ri axial ratios (Ri=2.0) and orienta-tions of φi lying between − 90° and + 90° to a reference line. On an R/φ diagram, the group of points representing this group of ellipses define a straight line passing through Ri = 2.0 (fig.II.7a). This group of ellipses can be made to undergo a characteristic strain in which the long axis is parallel to the reference line. For each stage shown on figure II.7 there are corresponding points for each ellipse on the Rf/φ diagram. If the strain ellipse has an Rd axial ratio such as Rd < Ri(fig.II.7b and c), the group of points on the Rf/φ diagram has a bell-shaped arrangement. The axial ratio of ellipse 10 in which the long axis is parallel to λG increases, meanwhile that of ellipse 1, where the long axis is parallel to λP decreases. For Rd = Ri, ellipse 1 is reduced to a circle Rf = 1. For values of Rd in which Rd > Ri, the Rf/φ diagram has a pear shape that becomes more and more elongated as Rd increases (fig.II.7 d and e). On these pear-shaped diagrams, the maximum (Rf max) and minimum (Rf min) values enable the axial ratio Rd of the deformation ellipse to be found:

Rd = Rfmax.Rfmin.

and the initial axial ratio of the ellipses :

Ri = Rfmax/Rfmin.

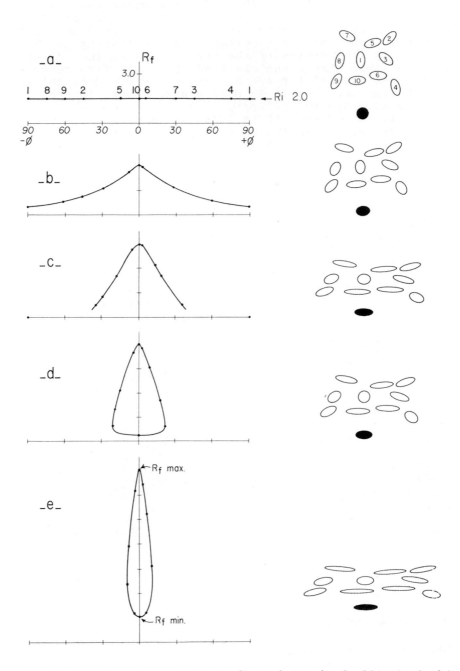

Fig.II.7. Progressive strain from a) to e) of elliptical objects which initially had a random orientation and corresponding Rf/φ curves of finite strain.

It is rare, nevertheless, for rocks to have elliptical objects so arranged that their axial ratios are constant. The construction of Rf/ϕ diagrams in actual cases generally gives a cloud of points and no simple and regular curves (fig.II.8). One can easily overcome this difficulty by looking for the best fit between the cloud of points obtained and a set of theoretical curves (fig.II.9) constructed for range of ratios between 1.5 and 4.0. Other problems are also posed in the application of this method to actual cases. The elliptical objects often have an initial preferred orientation, as in the case of pebbles in a conglomerate where the pebbles tend to lie flat in the plane of stratification. To obtain the correct value for Rd it is necessary to allow for this effect.

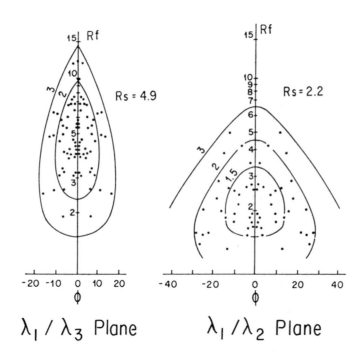

Fig.II.8. Application of the Rf/ϕ method to a natural case of metamorphic nodules (Le Theoff, 1977. Thesis, Rennes).

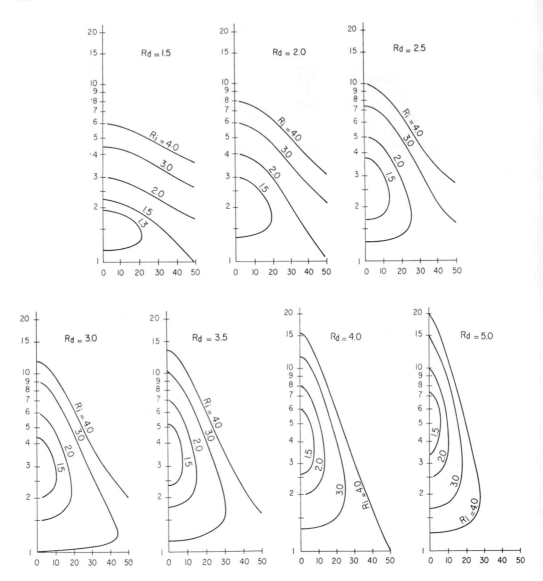

Fig.II.9. Rf/φ diagrams corresponding to Ri ratios of 1.5 to 5.

II.3.4. Variations in the distribution of point markers

        If one applies a homogeneous deformation to a sample in
which objects of equal diameter or points are arranged isotropically, the minimum distance between the objects changes, with
that along λ P getting decreasing and that along λ G increasing.
This fact is the foundation of the **centre to centre method**, which
is for example, applicable to a sedimentary rock containing
ooliths ; these initially spherical markers often undergo solution which excludes all direct strain measurement (fig.II.10).

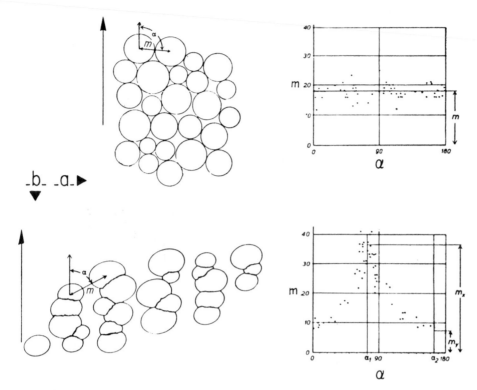

Fig.II.10. "Centre to centre" method of measurement of finite strain. Distribution of ooliths and corresponding m/α diagram : a) in the undeformed state and b) in a deformed state (Ramsay, 1967. McGraw Hill, New York).

In practice, one measures two by two the distance m between the centres of the objects under consideration on a surface $\lambda i \lambda j$, following directions related to a chosen reference in the plane of measurement (angle α). From an m/α diagram (fig.II.10), one can obtain an α1 direction in which, statistically, the distances between the centres are at a maximum mx and a direction α2 in which the distances are at a minimum my ; α1 is the angle between λ G and the chosen reference ; α2, the angle between λ P and this reference ; and the ratio mx/my, the ratio λ G/λ P.

This principle led Fry (1979) to propose a more general method (**Fry's method**) which is applicable in the case of an initially homogeneous and isotropic distribution of point markers which did not interact during the course of deformation. Thanks to a graphical procedure (fig.II.11), one can see the average distance between each marker for different orientations of the plane. In fig.II.11a, the distribution of initially isotropic point objects are considered : the smallest distance between the objects is nearly equal whatever the direction along which this distance is measured. After deformation by simple shear of γ = 1, then γ =2 (fig.II.11b and c), these minimum distances increase in the λ G direction and diminish in the λ P direction. The resulting ellipse is the strain ellipse in the plane under considera-tion.

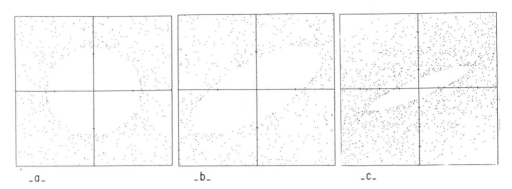

Fig.II.11. Fry's method of measuring finite deformation. a) Initial state. b) Simple shear with γ =1. c) Simple shear with γ =2. The method consists of successively placing each marker point at the centre of the diagram and plotting all the others. The empty central zone delineates the strain ellipse.

This method only applies in the following conditions :
  i) The minimum number of objects counted is about 100.
  ii) The deformation is homogeneous at the scale of the minimum distance between the objects.
  iii) There are a constant number of objects present during the course of deformation, thus the mineral objects must not increase (nucleation and growth) nor divide (fracturing).

II.4. DETERMINATION OF THE STRAIN ELLIPSOID

II.4.1. Measurement of ellipses

Usually, one defines the strain ellipsoid by starting with the determination of one or more ellipses. If only one strain ellipse is known, as when exposed fossils lie in the plane of the stratification, one also needs to know the orientation of , the principal axes of the ellipsoid. The ellipse under consideration must not contain any of the principal axes of the ellipsoid to obtain a solution. Most often one tries to find the ellipse in at least two of the three principal strain planes by selecting planes parallel to the $\lambda i \lambda j$ planes from measurement in the field. This operation is more easily carried out on intensely deformed rocks, where the lineation due to stretching is parallel to $\lambda 1$, and the plane of schistosity is parallel to the $\lambda 1 \lambda 2$ principal plane. Otherwise the measurement can be carried out upon properly sawed sections.
    There are also analytical methods for estimating the strain from measurements in three sections of known orientation. These methods use the principles of re-orientation of plano-linear markers and the geometry of invariant surfaces.

II.4.2 Reorientation of plano-linear markers

Let us consider a plane which before deformation is oblique to the future principal strain axes (fig.II.12). The angles which this plane makes with $\lambda 1$, $\lambda 2$ and $\lambda 3$ in the principal planes $\lambda 1 \lambda 3$, $\lambda 1 \lambda 2$ and $\lambda 2 \lambda 3$ are designated as $\theta 1$, $\theta 2$, $\theta 3$. After deformation these angles become respectively $\theta' 1$, $\theta' 2$ and $\theta' 3$, defined as follows :

$$\tan \theta' 1 = \lambda 3 / \lambda 1 \tan \theta 1$$
$$\tan \theta' 2 = \lambda 2 / \lambda 1 \tan \theta 2$$
$$\tan \theta' 3 = \lambda 3 / \lambda 2 \tan \theta 3$$

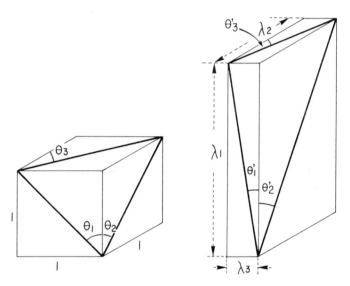

Fig.II.12.   Rotation of an arbitrarily oriented plane during deformation.

These equations show that the change in orientation of the plane is independent of the absolute value of the principal elongations and only depends upon their ratios. From these equations, it is possible to calculate the initial orientation of any plane knowing the axial ratios of the strain ellipsoid. Inversely, if one knew the initial and final orientations of a plane, one could calculate the axial ratios of the strain ellipsoid. In practice, this is not possible except in a statistical fashion. If we consider an initial distribution of planes in space as being uniform, after deformation there will be a preferred distribution of planes whose symmetry will enable one to calculate the axial ratios of the strain ellipsoid. Figure II.13 shows the preferred orientations of planes and lines, having an initially uniform distribution in space in a Flinn diagram.

This analysis can be applied to the shape preferred orientations ("fabrics") of rocks which contain minerals which are thought to have been passively re-orientated by strain (see also §7.3). Thus the analysis of preferred planar orientations of

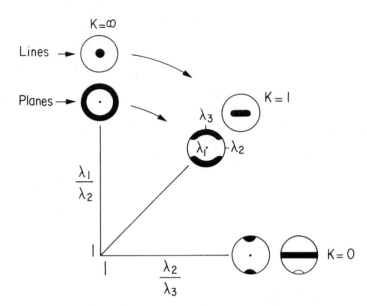

Fig.II.13. Flinn diagram showing preferred orientations of lines and planes of an initially random arrangement in cyclospherical projection ($\lambda 1$: vertical ; $\lambda 2$: E-W and $\lambda 3$: N-S).

micas or plano-linear orientations of amphiboles enables one to calculate a "fabric ellipsoid". Studies in which the fabric ellipsoid and the strain ellipsoid in the same rock have been compared, show that the two types of ellipsoid are only comparable when strain is moderate. The fabric ellipsoid remains coaxial with the strain ellipsoid but reflects imperfectly the strain magnitude (§6.2.5 and 6.3.5).

II.4.3. Deformation of mineral veins

Many rocks contain mineral veins. In a medium which has deformed in a homogeneous manner these veins are often deformed, sometimes by folding, sometimes by boudinage, sometimes by a combination of these two, as a result of their greater competence than that of their matrix. The type of structures that develop in a vein depends upon the orientation of the vein to the principal strain axes and to the symmetry of strain ellipsoid, that is to say to their corresponding elliptical section. Fig.II.14 shows a classification of ellipses which represents diagrammatically typical structures of mineral veins that have undergone the two main types of deformation represented by the strain ellipsoid : flattening and constriction. As folded and boudinaged veins are formed in both cases but for different orientations in relation to the principal strain axes, only doubly folded or boudinaged veins are characteristic of the type of strain ellipsoid.
    If there is a sufficient range of orientations in a rock which contains a large number of veins which are doubly folded (by constriction) or doubly boudinaged (by flattening), a

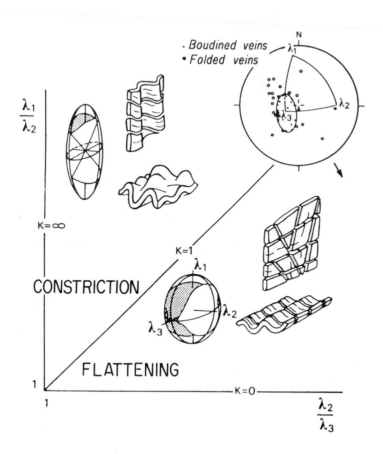

Fig.II.14.   Types   of   finite strain ellipsoid found by means   of
invariant   surfaces.   Projection   of   folded   and/or   boudinaged
quartzo-feldspathic   veins in a gneiss.   A flattening strain   has
been found (Brun, 1981. Thesis, Rennes).

cyclospherical   projection can be made in which their   respective
fields are delineated (fig.II.14). Thus are defined the invariant
surfaces (Talbot,   1970).   The angles between the invariant   surfaces   are measured in the two principal planes which they inter-
sect   , ψ 13   and ψ 23 for flattening and ψ 12   and ψ 13   for
constriction.   The   angles enable the axial ratios of the   strain
ellipsoid to be calculated (fig.II.15).
    This   simple   method   is above all applicable to the   case   of
flattening,   in   which   far fewer measurements are   necessary   to
correctly   define   the   invariant surface than   in   the   case   of
constriction. This method is very sensitive to changes of volume.
Within   these   limits   it is very useful in the   field   where   it
enables a rapid estimate of the strain ellipsoid to be made.

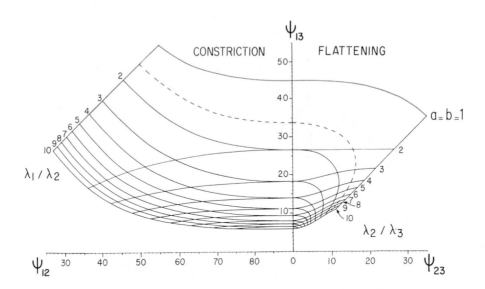

Fig.II.15.   Diagram   relating   the angles between   the   invariant
surfaces   and   the   axial ratios of the strain   ellipsoid   (after
Talbot, 1970).

II.5. STRAIN IN SIMPLE SHEAR

Shear zones are very common natural structures (§8.3). Although deformation is heterogeneous and can depart from a plane
regime, it can be broken down for analysis into the sum of elements deformed in simple shear with shear angles which increase
towards the centre of the shear zone (fig.8.12). From a sum of
the strain achieved by the simple shear elements, one can theoretically deduce the amount of relative displacement of the domains
adjacent to a shear zone. An empirical formula has also been
proposed which relates the width of a shear zone L to the displacement D (Otsuki, 1978) :

Log D = 1.01 log L + 1.78

According to this formula, simple shear in shear zones tends
towards a value of $\gamma \sim$ 60. This suggests that strain in a shear
zone can be considerable. It is therefore important to look for
strain markers.

It is possible to measure shear strain if the amount of
rotation of planar markers can be measured in the (X,Z) plane
The two main cases that can be considered are firstly a situation
where the shear plane has been located, to which one can relate
the orientation of induced foliation or some other planar markers, and secondly a situation where the shear plane has not been
identified but the amount of rotation of microstructural markers
can be measured.

II.5.1. **Shear plane/foliation obliquity**

We have seen that in simple shear, the principal axes of the shear ellipse initially oriented at 45° to the shear plane, gradually rotate, the long axis approaching the shear plane and that the measurement of the angle α enables one to determine γ (§II.2.2). We have also seen that the cleavage or foliation caused by this deformation generally coincides with the plane of flattening of the strain ellipsoid (§6.2.5). Thus, the trace of the foliation follows the path of the X axis of the strain ellipsoid (fig.8.12). Indeed, in shear zones, the foliation lies progressively nearer to the shear plane as the amount of the strain increases approaching the centre of the zone (§8.2).

Each time that the orientation of the shear plane can be identified and provided that the cleavage or foliation coincides with XY, γ can be measured starting from : $\gamma = 2\cot 2\alpha$. Figure II.16 illustrates this method.

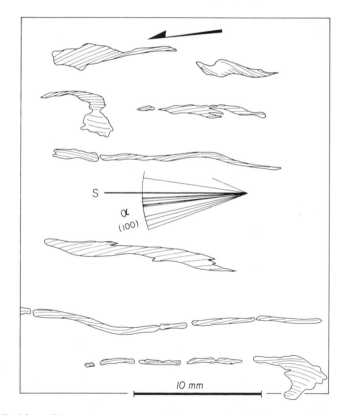

Fig.II.16. Shear measurement in a mylonitic peridotite using the obliquity between the foliatin S and the trace of the (100) slip planes of orthopyroxenes (thin section in a plane perpendicular to S and parallel to L). The smallest angle α is 2.5°, corresponding to $\gamma = 23$; the average angle is 9.5°, giving $\gamma = 6$. These values are under-estimates as the orthopyroxene is not as ductile as the olivine of the matrix as shown by the boudinaged strips of pyroxene. Note the two examples of reversed shearing and the retort shape of some pyroxenes (fig.8.18h).

When a shear zone is well exposed in the field, the shear plane is identifiable immediately as being the median plane of greatest deformation (fig.8.11). It is also locally seen as C surfaces (fig.8.13). On the other hand, in the case of a homogeneous deformation, the only way in which this plane can be identified is by analysis of the lattice preferred orientations of minerals that have been plastically deformed (§7.5.3). However, the estimate of γ from the angle α between the shape and lattice fabrics does not always coincide with that which is deducible from other markers (for example through the formula $\gamma = \lambda 1 - \lambda 2 = 2 \cotan 2\alpha$). This lack of agreement may have several causes : an identified component of coaxial flattening, the effects of subordinate slip systems upon the lattice preferred orientation, imprecision in measurements, etc. As a result this method of quantifying strain must be used with caution.

II.5.2. Rotation of planar markers

If the initial orientations of a planar marker, such as a vein, and that of the shear plane are known, γ can be found from the following formula (fig.II.17) :

$$\gamma = \cotan \beta' - \cotan \beta$$

where β is the initial angle between the vein and the direction of shearing, measured from outside the zone that has been affected by the shear for example, and β' is this angle measured from inside this zone. If $180° > \beta > 90°$, the rotation accompanying shearing has caused shortening and eventually folding which is detrimental to the measurement of β'.

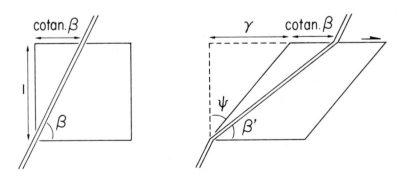

Fig.II.17. Rotation of a planar marker in simple shear.

II.5.3. Rotation of microstructural markers

The amount of rotation that a rigid object has undergone in the course of shearing can be recorded by crystallization in sheltered zones (§8.2.1) and in a growing porphyroblast by snowball or helicitic inclusions (§8.2.2). The amount of rotation gives an average strain estimate. This method is all the more

interesting as the corresponding markers are rather common and in
contrast to other methods, it records large deformations with a
theoretically constant precision.

Figure II.18 illustrates the ideal case of an undeformable
circle caught up in shear without relative slip between the
matrix and the circle. The derived formula $\gamma = \omega/2$ theoretically
allows one to calculate the shear strain from the amount of
rotation ω undergone by a marker attached to this circle.

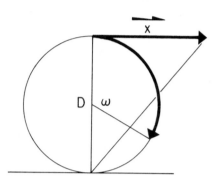

Fig.II.18. Relationship between γ and the rotation ω of a rigid
circle. For a displacement of x, $\gamma = x/D$ and also $x/\pi D = \omega/2\pi$
from which ω(radians) = 2γ .

In practice, one can attempt to "undeform" the deformed
structure in the case of pressure fringes in sheltered zones
(fig.8.4.b) ; for spiral inclusions, one measures the angle ω
between the orientation of the inclusions in the centre of the
porphyroblast and that of the foliation of the matrix, outside
the zone of perturbation caused by the porphyroblast (fig.II.19).

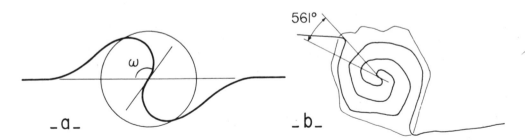

Fig.II.19. Measurement of the angle ω ;a) $\omega = 120°$, b) $\omega=561°$
(b) Rosenfeld, 1970).

In both cases there is a risk of underestimating the true value
of ω as the period of growth of fibres or of the porphyroblast
may not cover entirely that of the deformation ; also, the hard
object may not entirely record the amount of rotation due to its
surface slipping in relation to the matrix. Finally, in the case
of helicitic inclusions, the angle ω only measures the true

amount   of rotation if the schistosity recorded by the inclusions
in the centre of the porphyroblast was originally parallel to its
actual orientation in the matrix.   If the schistosity was   formed
at the same time as the shearing,   its initial orientation was at
45° to the shear plane and its final orientation makes an   angle
 α with this plane. The angle ω must be reduced by 45° $-\alpha$.
Situations   can   be   imagined where this effect could lead   to   a
spiral rotation in the opposite sense (Dixon, 1976). This type of
marker   gives   a   better result when the amount of   rotation   is
large.

FOR FURTHER READING

Brun, J.P., 1981. Instabilités gravitaire et déformation de la croûte
continentale. Thèse, Rennes, 197p.
Dixon, J.M.,1976. Apparent "double rotation" of porphyroblasts during a
single progressive deformation. Tectonophysics,34,101-115.
Dunnet,D.,1969.A technique of finite strain analysis using elliptical
particles. Tectonophysics, 7,117-136.
Fry,N.,1979. Random points distributions and strain measurements in
rocks. Tectonophysics,60,89-105.
Le Théoff, 1977. Marqueurs ellipsoidaux et déformation finie, Thèse
Rennes, 96 p.
Nadai,A. 1963. Theory of flow and fracture of solids. McGraw Hill éd.New
York,705 p.
Otsuki,K.,1978. On the relationship between the width of shear zone and
the displacement along fault. Jour.Geol.Soc.Japan,84,661-669.
Ramsay,J.G.,1967. Folding and fracturing of rocks. McGraw Hill éd.,New
York,568 p.
Rosenfeld,   J.L.,1970.   Rotated   garnets   in   metamorphic   rocks.
Geol.Soc.Am.Sp.Pap.,129,105 p.
Talbot,C.J.,1970. The minimum strain ellipsoid using deformed quartz
veins. Tectonophysics,9,47-76.

Appendix III

Cyclospherical Projections and Figures

III.1. INTRODUCTION

When the structural elements are expressed by the orientations of planes and lines, it is very useful to be able to project these on a plane for their analysis. The simplest planar representation for representing foliations is a structural map showing directions and dip by symbols. There is a large number of other ways of representing these, depending upon the nature of the problem posed : thus rose diagrams and histograms are well suited to the study of the distribution of a population of directions contained in a plane, or maps of axial distribution analysis (A.D.A. or A.V.A. "Achsenverteilung analyse") to represent the preferred orientations of domains thanks to a choice of captions corresponding to different angular sectors.

If the domain under study is statistically homogeneous, a cartographical representation, such as A.D.A., is not necessary and one can use representational techniques that take no account of the **location** of the object but only of its **orientation**. As they are most commonly used, the only projections presented concisely here are cyclospherical ones. There are several works that give details of their use (see the end of this appendix).

III.2. DEFINITIONS

Directions in space are considered irrespective of their location, each being represented as passing through the centre of a sphere of projection. A given direction is thus represented by its point of intersection with the hemisphere which has been chosen for the projection. A **cyclospherical projection** relates bi-univocally a point on a hemisphere to a point on its equatorial plane which is the plane of projection. The cyclopherical projection of the direction considered is finally that of its representative point on the sphere as plotted on the plane of projection (fig.III.1). Although this choice is arbitrary, following common practice the projections are plotted here as starting from the lower hemisphere as in figure III.1.

There are different projection modes of which we shall study the two most common ; the **equiangular** projection, also called **stereographic** and the **equal-area** or **Lambert** projection. Figures III.1a and b illustrate respectively the principles used in equiangular and equal-area projections. In the case of equiangular projection one can see that the projection of the equatorial circle of the sphere is this very circle which is not the case for the equal-area projection.

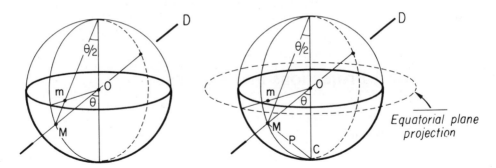

Fig.III.1. Principle of projection of the lower hemisphere upon the equatorial plane. The structural line D passes through O cutting the lower hemisphere (of radius R) at M ; m is the projection of M in the equatorial plane, such that : a) in equiangular projection, Om = Rtan(θ /2), b) in equal-area projection Om = MC = 2R sin (θ/2).

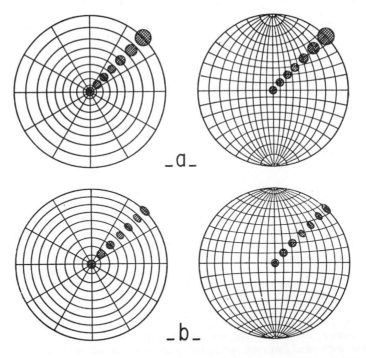

Fig. III.2. Polar nets on the left and meridian nets on the right correspond to the two principal types of projection : a) equiangular ; b) equal-area. The dotted zones show that in the equiangular projection (a) the angles are preserved (the circular form is retained) but not the areas. In the equal-area projection (b) on the contrary, the areas but not the angles are preserved (circles become ellipses) (Vistelius, 1966. Pergamon, Oxford).

If a system of reference lines is drawn upon the hemisphere, for example meridional and parallel (i.e. longitudinal and latitudinal) their projection upon the plane forms a **network or net**. The nets are used for plotting and reading the orientations of the directions being studied. For each type of projection one can use different nets (fig.III.2). Thus according to whether the line of poles of the system of coordinates on the sphere coincides with the plane of projection, or on the contrary with its normal, the net is **meridional or polar**. The polar net is hardly used and only for illustrating structures with axial symmetry.

Examination of a terrestrial map shows that projections introduce distortions into the representation of geographical data. Equiangular projection retains the angles and equal-area projection, the areas (fig.III.2). The first is traditionally used, with the corresponding meridional net or **Wulff net** (see plate at end of book), in problems dealing with the angular relations of the given structural elements, and the second, with the **Schmidt net** (see plate at end of book) is used in problems of a statistical type where numerous data lead to studies on the density of distribution of directions (fig.III.3). The Schmidt net which equally allows for description of angular relationships, is thus adequate in structural studies.

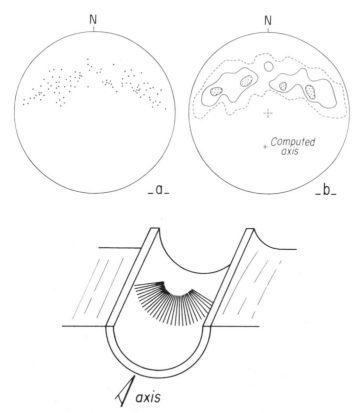

Fig.III.3. Equal-area projections (lower hemisphere) of poles to a folded surface. The poles are arranged in a girdle (great circle). a) Raw data (100 measurements). b) Dimitrijevic density diagram, contours at 1 4,8 % for 1 % of the net area.

III.3. REPRESENTATION OF STRAIGHT LINES AND PLANES

The representation of a straight line appears as a point (fig.III.1) on the net. A plane characterized by its normal is also represented by a point, called the **pole** of the plane ; one then speaks of a polar representation of a plane (fig.III.4). One can also project the plane directly ; its intersection with the sphere is a diametrical circle of which a point by point projection on the network leaves a **cyclographical** trace (fig.III.4), also called a **great circle** (meridian of the net) due to its correspondence with a diametrical circle on the sphere.

If the data have been measured and are projected with reference to the horizontal plane and the N-S and E-W coordinates, the most commonly met situation in structural geology, the cyclographic trace of a vertical plane is a diameter of the net and that of a horizontal plane, the circle limiting the net (fig.III.5). If the chosen system of reference is not the geographical one, one can adopt another plane of projection. Thus in **petrofabric analysis** (study of preferred orientation of minerals in thin sections) the plane of reference often is that of the thin section itself to which the plane of projection is itself parallel. One can change the reference plane (by rotation) in order to present the fabric data in another plane of reference.

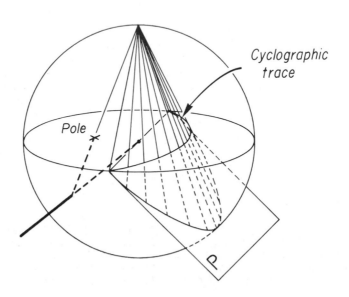

Fig.III.4. Polar (x) and cyclographic (line) projections of the plane P (equiangular projection).

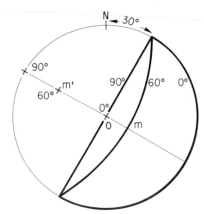

Fig.III.5.   Polar (x) and cyclographic (thick lines) projections
of planes   at   North 30° East and dipping at   0°-60°-90°.   The
projection being equiangular,   the distance Om which corresponds
to   the   line   of greatest dip in the plane dipping at 60° is
Rtanθ/2 = 0.26R,   with θ   = π/2 - 60° (fig.III.1a). The distance
of pole m' from this plane to the centre is:Om'=Rtan60°/2=0.58 R.

Direct and inverse representations. A structural object is completely described by three vectors defining its internal reference frame.   One orientates it by making a correlation between this
reference frame and an external and invariant reference frame ;
one   thus   orientates a mineral by three crystallographic axes
(internal references) in locating each of these axes in relation
to an external reference,   which is usually that of the sample,
defined by the foliation and lineation (fig.7.18b). As the orientation of the crystal is given in relation to this external
reference,   the   representation is **direct**.   It is **inverse** if the
chosen reference is that of the crystal to which are then related
the directions attached to the external reference ; thus in the
case of an experimental uniaxial deformation of a polycrystalline
aggregate, the direct mode consists of plotting the crystallographic axes by reference to the stress σ1 (fig.III.6a) and the
inverse mode, by plotting the stress in the crystallographic
reference (fig.III.6b).   The inverse method is little used except
in searching for specific orientation correlations,   such as evidence of crystallographic slip systems.
     A   common situation is that of the direct representation in
the geographical coordinates,   of foliations and lineations
constituting   the   structural   reference   frame of the sample
(fig.III.7a and b).   In petrofabric analysis, a correlation is
sought between the orientations of the crystallographic axes of
the minerals of a sample and its foliation-lineation system which
is preferably taken as the coordinates of the net (fig.7.18b).
The   latter system must be completely defined,   namely the lineation must be represented in the foliation plane.

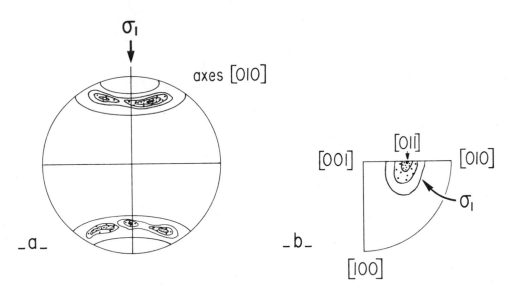

Fig.III.6. Diagrams showing a) the direct and b) the inverse
representations of the relationship between the applied uniaxial
stress and the crystallographic directions of a deformed
monophase aggregate. In the direct case the reference direction
is the principal stress and the orientations of the crystallographic axes are related to it. On the contrary in the inverse
diagram the reference directions are the crystallographic axes
and the orientation of the principal stress measured with respect
to each crystal is plotted (the symmetry of the system only
necessitates the representation of a sector of the diagram). The
inverse representation immediately shows the coincidence of the
direction of stress and the crystallographic axis [011].

III.4. DENSITY DIAGRAMS

In what follows, we suppose that the collected data are
representative of the phenomenon studied, although in structural
geology this question may be far from trivial due to the limited
and discontinuous nature of measurement sites in relation to the
size and complexity of natural structures.
The problem posed being that of the study of a large number
of measurements of directions expressed by a cloud of points on a
net, one looks for a new representation more suited to visual
analysis (fig.III.3). The method aimed at is to dilute the point
information as uniformly as possible over the whole net. The
techniques used consist of substituting for clouds of points,
areas of equal density per unit area (by means of a counting cell
of constant area), separated by density contours. The corresponding diagram is called a **diagram of orientation density** or a
density diagram. One does it by moving the counting cell in order
to cover the area of the sphere or of the net completely, in
making it coincide either with pre-established nodes of a network
or with the points of measurement themselves.

There   are many manual or automatic techniques.   The   manual
techniques   use cells drawn on the net,   which can have a square,
circular   or elliptical shape.   This last shape results from   the
equal-area   projection of spherical counting cells ;   it has   the
merit   of limiting otherwise considerable   spherical   distortions
(fig.III.2). The corresponding net is that of Dimitrijevic (plate
at the end of the book). The techniques calling upon the use of a
computer   allow a direct count upon surface of the sphere,   starting with   spherical cells.   The most elaborate seems to be   that
of   Bouchez and Mercier which uses a highly symmetrical   counting
network.   In manual counting the density is expressed in % of 1 %
of the area of the net.   In Bouchez and Mercier's technique,   the
area   of the cell is determined as that which minimises the   gaps
and overlaps made in the counting grid.   The area of this cell is
then   only 0.45 % of the surface of the hemisphere.   Figure 7.18b
is an illustration of a density diagram obtained by this   method.
The   use of a large area counting cell introduces a corresponding
smoothing of contours which,   in some cases,   can cause a loss of
information,   or contrarily,   result in extracting a   significant
but diffuse information.

III.5. **ANALYSIS OF DENSITY DIAGRAMS**

When   the   analysed structure is homogeneous or has   a   high
degree   of   symmetry,   as in the case of conical   or   cylindrical
folds, density diagrams reflect this symmetry and can be analysed
in terms of a restricted number of patterns, possibly associated.
The following patterns can be distinguished :
1)   The   **point maximum** for which a **best axis** (b.a.)   can   be
computed (fig.III.7a).   This distribution corresponds to a bundle
of directions which are almost parallel.
2)   **The   girdle**,   in which the points are grouped   around   a
great circle of the net (projection of a plane) from which a **best
pole** (b.p.) can be computed (fig.III.3 and 7b). This distribution
corresponds   to a preferred orientation of the data points   close
to a plane.
3)   The   **small circle**,   for which the corresponding   straight
lines   are preferentially distributed upon the surface of a   cone
wich is defined by its axis (b.a.) and by its opening (fig.III.6a
and III.7c).

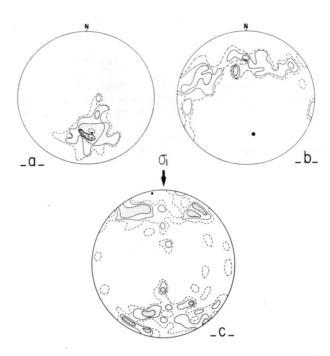

Fig.III.7. Principal types of distribution in density diagrams (equal-area projections, lower hemisphere ; about 100 measurements ; contours at 1,2,4,8 % for 0.45 % area) (Bouchez, 1977). a) Point distribution with best axis (dot). b) Girdle distribution with best pole (dot). These diagrams relate to a fold, with in b) the poles of the folded surface defining the fold axis and in a) the mineral lineations parallel to the fold axis. c) Small circle distribution with the corresponding cone axis (dot) in an experimental deformation showing the c crystallographic axes of quartz.

FOR FURTHER READING

Bouchez, J.L., 1977. Traitement automatique des données directionnelles. Trav. Lab. Tectonophysique Nantes.

Nicolas, A. et Poirier, J.P., 1976. Crystalline plasticity and solid state flow in metamorphic rocks. Wiley-Interscience, London.

Ragan, D.M., 1973. Structural geology. An introduction to geometrical techniques. Wiley, New York.

Turner, F.J. et Weiss, L.E., 1963. Structural analysis of metamorphic tectonites. McGraw Hill, New York.

Vialon, P., Ruhland, M. et Grolier, J., 1976. Eléments de tectonique analytique. Masson, Paris.

Vistelius, A.B., 1966. Structural diagrams. Pergamon, Oxford.

Winchell, A.N., 1961. Elements of optical mineralogy, Part I. J. Wiley, New York, p.239-254 (introduction à l'usage de la platine universelle).

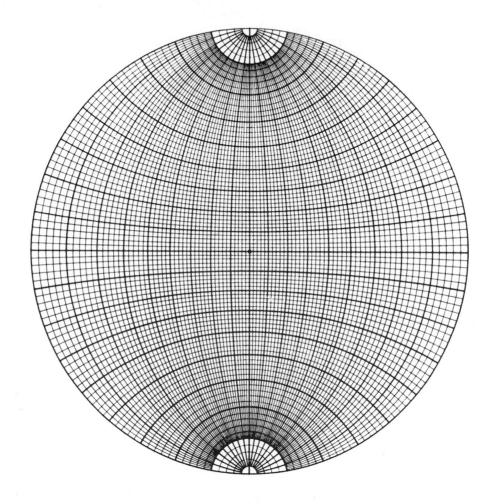

Wulff net

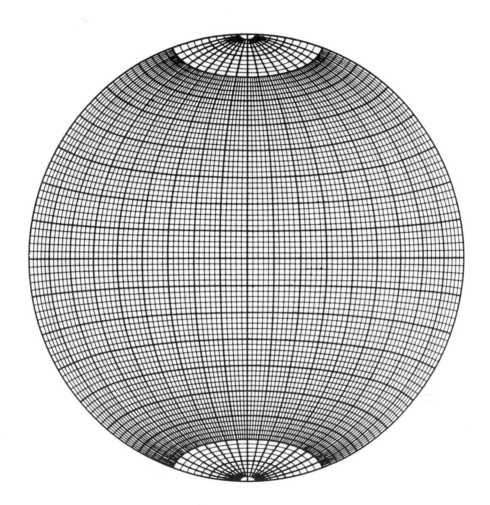

Schmidt net

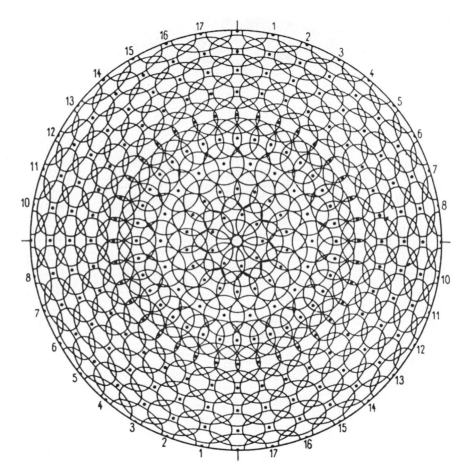

Dimitrijevic net

Index